TEST BANK

to accompany

Computers and Data Processing

TEST BANK

to accompany

COMPUTERS

DATA

AND

PROCESSING

HARVEY M. DEITEL
Boston College
BARBARA DEITEL

ACADEMIC PRESS, INC.
(Harcourt Brace Jovanovich, Publishers)
Orlando San Diego San Francisco
New York London Toronto Montreal
Sydney Tokyo São Paulo

Academic Press, Inc.
Orlando, Florida 32887

United Kingdom Edition Published by Academic Press, Inc.
(London) Ltd., 24/28 Oval Road, London NW1 7DX

ISBN: 0-12-209023-3
Printed in the United States of America

Table of Contents

Preface vii

Part One: Introduction

1. The Information Revolution 1
2. The Evolution of Computers 17

Part Two: Hardware

3. The Processor 39
4. Input: Gateway to the Computer 64
5. Output: Getting Results from the Computer 86
6. Secondary Storage 107
7. Data Communications 126

Part Three: Software

8. Structured Programming 156
9. Programming Languages 179
10. Structured Systems Analysis and Design; Systems Acquisition 205
11. Database Management Systems, Management Information Systems, Decision Support Systems 244
12. Operating Systems 270

Part Four: Computers in Business

13. Personal Computing 307
14. Office Automation 328
15. Electronic Funds Transfer Systems, Security, Privacy, and Computer Crime 353

Part Five: Computers in Society

16. Robotics and Artificial Intelligence 378
17. Computers and Medicine 401
18. Computers and the Handicapped 413
19. Computers and Transportation 431

Preface

Thank you for adopting our text, ***Computers and Data Processing***, for use in your introductory course. This Test Bank has been prepared to assist you in the important task of developing thorough and meaningful quizzes, tests, and final examinations.

The objectives of this Test Bank are clear: (1) Provide a broad variety of questions from which instructors can easily custom tailor exams appropriate for their particular courses. (2) Provide a very large number of questions, many per section and some similar to one another, so that instructors can produce different exams testing essentially the same topics in succeeding semesters. (3) List the answers side-by-side with the questions so that instructors can rapidly locate questions that are appropriate and unambiguous in the context of their courses. (4) Group all questions by section heading within the main text to enable instructors to select questions by section and by enrichment piece.

The Test Bank contains over 4000 short-answer questions that span the full range of regular topics and enrichment pieces in ***Computers and Data Processing***. This is approximately two to three times as many questions as are normally provided in test banks for introductory computing courses. Three types of questions are included: multiple choice, true/false, and fill-in-the-blanks.

Answers for the questions are provided in-line with the questions rather than requiring you to flip to separate answer keys. We suggest that you read each question and answer to see if it fits the context of your particular course. A computerized testing service is being provided by Academic Press. Each of the 4000 questions is available on disk; software is provided to help you select and edit the questions you wish to include.

Many instructors prefer to use multiple-choice and true/false questions to form machine-gradable exams, especially when large course sections are involved; approximately 2800 such questions have been provided. Some instructors feel that fill-in-the-blanks questions are appropriate because they force the student to read the material very carefully; approximately 1200 fill-ins have been provided.

For each chapter, the questions of each type are organized by the headings under which they appear in the main text. Instructors will find this helpful in localizing questions to those sections and enrichment pieces actually covered in their courses.

Answers for the multiple choice questions are indicated by an asterisk (*) several spaces to the right of the correct choice. Answers for the true/false questions are indicated in parentheses after the statement; the page number of the main text on which each answer may be found is also listed. Answers for the fill-ins are shown in boldface text in the statements themselves.

We would sincerely appreciate your comments and criticisms of this Test Bank and other course materials as well. We will repond to all correspondence immediately. Please address all correspondence to:

Harvey and Barbara Deitel (Authors)
c/o Computer Science Editor
Academic Press, Inc.
Orlando, FL 32887

Chapter 1
The Information Revolution

Multiple Choice

Introducing the Computer

1. The _____ Revolution is a period of change that has been caused by advances in computer technology.
   (a) Microprocessor
   (b) Software
   (c) Information *
   (d) Computer
   (e) Industrial

2. Computers help us to extend our _____ power.
   (a) brain *
   (b) user-friendly
   (c) microprocessor
   (d) muscle
   (e) hardware

3. Due to the vast amounts of data we process and use we have become a(n) _____-based society.
   (a) software
   (b) silicon chip
   (c) hardware
   (d) data communications
   (e) information *

4. Computerized _____ are replacing many blue-collar workers on assembly lines.
   (a) lasers
   (b) robots *
   (c) programmers
   (d) expert systems
   (e) scanning systems

5. _____ are tiny computers etched onto thin slivers of silicon called silicon chips.
   (a) Computer programs
   (b) Lasers
   (c) Computer graphics

(d) Microprocessors *
(e) Operating systems

6. The instructions that computers follow are called computer ______.
(a) outputs
(b) graphics
(c) hardware
(d) flowcharts
(e) programs *

7. The software that a computer uses is written by computer ______.
(a) hardware
(b) outputs
(c) programmers *
(d) graphics
(e) expert systems

8. Many household appliances have ______ computers.
(a) talking *
(b) laser
(c) fiber optic
(d) word processing
(e) intelligent

9. ______ cost a few hundred to a few thousand dollars, can fit on a desktop, and have the power of the "giant brains" of a few decades ago.
(a) Computerized scanning systems
(b) Fiber optics
(c) Computer programmers
(d) Personal computers *
(e) Lasers

10. Computerized ______ systems are used to typeset books.
(a) laser-scanning
(b) photocomposition *
(c) graphics
(d) touch-sensing
(e) data communication

The Benefits

11. Computers are essentially ______ tools.
(a) living

(b) thinking
(c) problem-solving *
(d) reasoning
(e) communications

12. To benefit from a computer you must ask the right questions about a problem and be able to ______ that information correctly.
(a) facilitate
(b) reduce
(c) extend
(d) conserve
(e) manipulate *

13. Which of the following is not always considered a benefit computers can offer? ______
(a) reduce tedious work
(b) help the handicapped
(c) explore space
(d) increase leisure time
(e) replace human workers *

The Dangers

14. Which of the following would not generally considered a danger that computers could bring? ______
(a) Invasion of privacy
(b) Massive unemployment
(c) Relegate people to pushing buttons
(d) Exploring space *
(e) Displacing workers

Looking to the Future

15. Systems that are easily understood by untrained people are said to be ______.
(a) user-friendly *
(b) touch-sensitive
(c) personal computers
(d) expert systems
(e) information-based

16. Speech ______ and ______ systems can converse with their users in everyday English.
(a) printing, holography
(b) communications, data flow

(c) recognition, synthesis *
(d) software, program
(e) sensing, graphics

17. ______ devices use an intense beam of monochromatic light.
(a) Silicon
(b) Touch-sensing
(c) DBMS
(d) Word processing
(e) Laser *

18. An example of user-friendliness is ______.
(a) personal computer
(b) phoneme coding
(c) pseudocode
(d) touch sensing *
(e) laser holography

19. Speech synthesis and speech recognition are considered to be ______
(a) software
(b) data communications
(c) user-friendly *
(d) pseudocode
(e) laser holography

A Tour of the Text

20. The history of the computing field is divided into four ______, that are characterized by major innovations in electronics technology.
(a) concepts
(b) generations *
(c) operating systems
(d) programming languages
(e) microprocessors

21. ______ the actual computing equipment.
(a) Microprocessors are
(b) Software is
(c) Programs are
(d) Hardware is *
(e) Laser holography is

22. The ______ is also called the brain of the computer.

(a) processor *
(b) software
(c) secondary storage
(d) pseudocode
(e) programming language

23. A form of computer input that allows a user to point to something on the display screen is called ______
(a) fiber optics
(b) data communication
(c) touch sensing *
(d) computer graphics
(e) holography

24. ______ is the most popular way today to enter information into a computer.
(a) Speech synthesis
(b) Keyboard *
(c) Laser holography
(d) Speech recognition
(e) Touch sensing

25. The results we receive from computers are called ______
(a) input
(b) pseudocode
(c) transborder data flow
(d) automation
(e) output *

26. Printers and display screens are examples of ______ devices.
(a) software
(b) three-dimensional
(c) user-friendly
(d) touch sensing
(e) output *

27. Three-dimensional TV may eventually be built with the technique of ______.
(a) fiber optics
(b) laser holography *
(c) computer graphics
(d) laser printing
(e) transborder data flow

28. The computerized preparation and processing of pictures by computers is called computer ______

(a) recognition
(b) synthesis
(c) sensing
(d) graphics *
(e) optics

29. Magnetic tape and magnetic disks are examples of ______.
(a) secondary storage *
(b) information-based systems
(c) operating systems
(d) data communications
(e) outputs

30. ______ is primarily concerned with the movement of information between computer systems.
(a) Transborder data flow
(b) Fiber optics
(c) Data communications *
(d) Laser holography
(e) Pseudocode

31. ______ is primarily concerned with the transmission of information across international boundaries.
(a) Electronic funds transfer
(b) Fiber optics
(c) Flowcharting
(d) Data communications
(e) Transborder data flow *

32. The communications cable technology of the future is called ______.
(a) transborder data flow
(b) fiber optics *
(c) laser holography
(d) silicon
(e) computerized scanning

33. Two techniques that foster the development of high-quality computer programs are structured flowcharting and ______.
(a) decision support systems
(b) database management systems
(c) operating systems
(d) pseudocode *
(e) management information systems

34. Programmers use different ______ in writing computer programs.
(a) software
(b) hardware
(c) programming languages *
(d) secondary storage
(e) flowcharting

35. Early computers were programmed in ______ language.
(a) machine *
(b) query
(c) BASIC
(d) APL
(e) Ada

36. ______ concerns the development, installation, evaluation, and control of computer systems.
(a) DBMS
(b) Structured flowcharting
(c) DSS
(d) Pseudocode
(e) Structured systems analysis and design *

37. Which of the following is not essential in structured systems analysis and design? ______
(a) Installation
(b) Development
(c) Database *
(d) Control
(e) Evaluation

38. ______ systems control the secure storage and accessing of information that a business must process.
(a) Management information
(b) Decision support
(c) Office automation
(d) Word processing
(e) Database management *

39. Computerized systems that provide information on a regular schedule to help in planning, organizing, directing, and controlling are called ______.
(a) office automation systems
(b) management information systems *
(c) operating systems
(d) decision support systems
(e) database management systems

40. Interactive computer-based planning systems that support management are called ______
    (a) word processing systems
    (b) operating systems
    (c) office automation systems
    (d) decision support systems *
    (e) database management systems

41. VisiCalc is an example of a ______ system.
    (a) database management
    (b) operating
    (c) management information
    (d) office automation
    (e) decision support *

42. ______ systems are primarily concerned with managing the computer's hardware.
    (a) database management
    (b) operating *
    (c) decision support
    (d) office automation
    (e) management information

43. The ______ operating system was developed at Bell Laboratories.
    (a) UNIX *
    (b) MS-DOS
    (c) CP/M
    (d) XENIX
    (e) PC-DOS

44. ______ is the operating system normally supplied with the IBM Personal Computer.
    (a) CP/M
    (b) XENIX
    (c) UNIX
    (d) PC-DOS *
    (e) MS-DOS

45. The combination of computers and communications can bring information into people's homes with a service called ______.
    (a) datatex
    (b) infotex
    (c) videotex *
    (d) lasertex
    (e) microtex

46. Word processing is the leading technology in the field of ______.
(a) data communications
(b) office automation *
(c) decision support
(d) management information
(e) database management

47. The office automation technology that is primarily concerned with transmitting pictures is ______.
(a) electronic mail
(b) word processing
(c) voice mail
(d) facsimile *
(e) computer graphics

48. The storage and movement of money as electronic signals is called ______.
(a) APL
(b) RPG
(c) DSS
(d) MIS
(e) EFT *

49. Computers that are able to play chess are examples of ______.
(a) artificial intelligence *
(b) management information systems
(c) database management systems
(d) decision support systems
(e) computerized scanning systems

50. ______ systems contain the cumulative knowledge and rules of thumb in a given field.
(a) database management
(b) decision support
(c) expert *
(d) operating
(e) management information

51. Computerized ______ systems allow doctors to see inside the human body without surgery.
(a) fiber optics
(b) microprocessor
(c) expert
(d) scanning *
(e) operating

52. The Kurzweil Reading Machine uses ______ techniques to read books to the blind.

(a) speech recognition
(b) optical-scanning *
(c) fiber optics
(d) data communications
(e) laser holography

53. The use of computerized ______ can help students learn to drive a car or fly a plane.
(a) scanning systems
(b) expert systems
(c) simulators *
(d) decision support systems
(e) database management systems

54. The ______ system uses only zeros and ones.
(a) laser holography
(b) base zero
(c) base one
(d) silicon
(e) base two *

True/False

Introducing the Computer

1. Computer technology is at the root of the Information Revolution. (T p4)

2. Most machines help us extend our brain power. (F p4)

3. Computerized robots are already replacing blue-collar workers on assembly lines. (T p4)

4. The cost of microprocessors is generally decreasing. (T p5)

5. Some microprocessors are as powerful as the room-sized computers of a few decades ago. (T p5)

Looking to the Future

6. The "help" key makes computers more user-friendly. (T p7)

7. Today's optical computers will generally be replaced with electronic computers

by the year 2000. (F p7)

A Tour of the Text

8. The fifth generation of computers appeared in 1980. (F p8)

9. Computer hardware is the actual computing equipment. (T p8)

10. Today, most information is entered into computers with touch sensing. (F p8)

11. Speech synthesis is an input method in which the computer recognizes spoken commands. (F p8)

12. The results from computers are called outputs. (T p8)

13. Laser printing is the printing technology of the future. (T p8)

14. Phoneme coding is a speech recognition technique. (F p8)

15. Computer graphics is concerned with the computerized preparation and processing of pictures. (T p8)

16. Bar-coded information is read by using laser beams. (T p9)

17. Magnetic tape and magnetic disks are secondary storage devices. (T p10)

18. Optical disk is the communications cable technology of the future. (F p10)

19. Two techniques that foster the development of high-quality computer programs are structured flowcharting and pseudocode. (T p10)

20. Programming languages have evolved from the query languages of the 1940s to the machine languages of the 1980s. (F p10)

21. Fiber optic cable is made of extremely pure plastic. (F p11)

22. Space satellites used for data communications are in fixed orbits 2,230 miles above the earth. (F p11)

23. DBMS stands for "decision-based management systems." (F p12)

24. The type of DBMS receiving the most attention today is the relational DBMS. (T p12)

25. MIS stands for "management information systems." (T p12)

26. The key tasks of managers are planning, organizing, directing, and controlling. (T p12)

27. DSSs are interactive computer-based systems that support management decision-making activities. (T p12)

28. DSSs are especially useful in support of the planning function. (T p12)

29. Electronic worksheets brought MIS systems to the attention of personal computer users. (F p12)

30. Operating systems are the software systems that manage the computer's hardware and make it more "friendly" and usable to computer users. (T p12)

31. The UNIX operating system was developed by Microsoft. (F p12)

32. Two of the most important personal computers for the 1980s are Apple's PC and IBM's Macintosh. (F p13)

33. "Long haul" networks tie together computer workstations within the office. (F p13)

34. Facsimile is an office automation technology that involves the transmission of pictures. (T p13)

35. EFT systems involve the storage and movement of our money as electronic signals. (T p13)

36. It is generally considered to be unlikely that a chess-playing computer will ever beat the world's best human chess player. (F p13)

37. Robots are also known as steel-collar workers. (T p13)

38. Anthropomorphic robots are widely used in industry. (F p14)

39. Computerized scanning systems require exploratory surgery to see inside a patient. (F p14)

40. From the standpoint of handicapped people, computers are difficult to use and are not likely to improve their lives. (F p15)

41. Professor Jerrold Petrofsky of Wright State University experiments with computer-controlled stimulation of muscles to help paralyzed patients move. (T p16)

42. The Kurzweil Reading Machine was designed to teach elementary school children how to read

quickly and accurately. (F p16)

43. In the base two number system numbers are expressed as combinations of ones and twos. (F p17)

Fill-in-the-Blanks

Introducing the Computer

1. The instructions that computers follow are called computer programs or **software**.

2. **Microprocessors** are tiny computers etched onto silicon.

3. The slivers of silicon used in today's computers are called silicon **chips**.

4. People who write the instructions that computers follow are called **computer programmers**.

5. We are in the midst of what is commonly called the Information **Revolution**.

6. Computers help us to extend our **brain power** rather than our muscle power.

7. The power of the prohibitively expensive "giant brains" of the past is now available in desktop **personal** computers.

The Benefits

8. Computers are essentially **problem**-solving tools.

The Dangers

9. Many people fear that computers will displace humans causing **unemployment**.

10. One of the greatest fears people have about computers are that they will invade our **privacy**.

Looking to the Future

11. Computer systems that are easily used by untrained people are said to be **user-friendly**.

12. The costs of computing have been (increasing/**decreasing**).

13. Speech synthesis and **speech recognition** will help computers converse with their users in everyday English.

14. A **laser** is a device that creates an intense beam of monochromatic light.

A Tour of the Text

15. The actual computing equipment is referred to as computer **hardware**.

16. The "brain" of a computer is called its **processor**.

17. **Biochips** may make "living" computers possible.

18. The process of entering data into a computer is called **input**.

19. The most popular device for entering information into a computer is the **keyboard**.

20. The user-friendly input method in which the user points to a symbol on the display screen is called **touch sensing**.

21. **Speech recognition** allows a computer to recognize spoken commands.

22. The results that the computer produces are called **output.**

23. **Laser** printers use intense beams of light to produce letters, digits, and pictures.

24. Computer-controlled lasers produce three-dimensional pictures in a technique called **laser holography**.

25. Computerized speech generation is called **speech synthesis**.

26. Computer **graphics** is the computerized preparation and processing of pictures.

27. Magnetic tape and magnetic disk are examples of **secondary storage** devices.

28. **Data communications** is concerned with the movement of information between computer systems.

29. The transmission of information across international borders is called **transbroder data flow**.

30. **Fiber optics** is considered the communications cable technology of the future.

31. Beams of laser light are transmitted over glass "wires" with **fiber** optics.

32. Computers use sets of instructions called software or **computer programs** to help solve problems.

33. With techniques such as structured flowcharting and **pseudocode**, high-quality computer programs can be developed.

34. A programming group called the **chief programmer team** is effective in the development of small- and medium-size software systems.

35. BASIC, FORTRAN, COBOL, RPG, C, and Ada are examples of computer **programming languages**.

36. A **database management system (DBMS)** controls the secure storage and accessing of the large amounts of information that a business must process.

37. **Management information systems (MIS)**, are computerized systems that provide management with information on a regular basis.

38. Managers use **management information systems (MIS)** to help them perform their ket tasks of planning, organizating, directing, and controlling.

39. A **decision support system (DSS)** is an interactive computer-based system that supports management decision-making activities, especially the planning function.

40. VisiCalc, the electronic **worksheet**, is an example of a decision support system.

41. The software systems that manage the computer's hardware to make it more user-friendly are called **operating systems**.

42. UNIX, XENIX, and MS-DOS are examples of **operating systems**.

43. The version of the MS-DOS operating system supplied with the IBM Personal Computer is called **PC-DOS**.

44. The two most important personal computers of the 1980s are probably the IBM Personal Computer and the **Apple Macintosh**.

45. **Office automation** deals with the use of computers and communications in support of the information-processing functions in offices.

46. Computerized text manipulation is called **word processing**.

47. The office automation technology that concerns the transmission of still pictures is called **facsimile**.

48. The storage and movement of money as electrical signals is called **electronic funds transfer**.

49. **Expert systems** are computer systems that function at the levels of human experts in various fields.

50. **Robots** are also known as steel collar workers.

51. The base two number system, also called the **binary number system**, expresses all numbers as combinations of zeros and ones.

Chapter 2
The Evolution of Computers

Multiple Choice

Early Computing Devices

The Abacus

1. The computing device that uses beads to represent units, tens, hundreds, and thousands is the ______.
   (a) slide rule
   (b) simultaneous punching machine
   (c) abacus *
   (d) Stepped Reckoner
   (e) Arithmetic machine

2. When using an abacus, the beads positioned at the center bar are ______.
   (a) not calculated
   (b) calculated *
   (c) subtracted
   (d) multiplied
   (e) divided

Napier's Bones

3. The computing device that uses sticks marked in a fashion similar to mutiplication tables is called ______.
   (a) abacus
   (b) Stepped Reckoner
   (c) Jacquard loom
   (d) slide rule
   (e) Napier's bones *

4. The principles of ______ led to the development of the slide rule.
   (a) abacus
   (b) Boolean logic
   (c) Jacquard loom
   (d) Napier's Bones *
   (e) base two numbers

5. Edmund Gunter developed the ______ which became extinct with the introduction of the pocket calculator.

(a) Difference Engine
(b) Analytical Engine
(c) slide rule *
(d) Stepped Reckoner
(e) arithmetic machine

Pascal's Arithmetic Machine

6. The computing device that works in a similar fashion to an automobile odometer is the _______.
(a) arithmetic machine *
(b) Stepped reckoner
(c) slide rule
(d) Difference Engine
(e) Analytical Engine

7. _______ designed a calculating machine that used connected toothed gears.
(a) John Napier
(b) Lady Lovelace
(c) Seymour Cray
(d) Maurice Wilkes
(e) Blaise Pascal *

Leibniz and the Stepped Reckoner

8. ______ modified Pascal's machine so that multiplication and division could be done directly.
(a) George Boole
(b) Gottfried Leibniz *
(c) Joseph Marie Jacquard
(d) Augusta Byron
(e) John Vincent Atanasoff

9. The ______ used a movable carriage operating on wheels that performed multiplication as a series of additions.
(a) Jacquard loom
(b) arithmetic machine
(c) Stepped Reckoner *
(d) Difference Engine
(e) Analytical Engine

Jacquard's Loom

10. The ______ was the first machine to use punched-card control.
   (a) Difference Engine
   (b) Analytical Engine
   (c) Stepped Reckoner
   (d) Jacquard Loom *
   (e) Arithmetic Machine

11. ______ was the first to use punched-card control.
   (a) Joseph Marie Jacquard *
   (b) Charles Babbage
   (c) George Boole
   (d) John Napier
   (e) John Billings

Babbage's Analytical Engine

12. The ______ was designed by Charles Babbage while he was a student.
   (a) Arithmetic machine
   (b) Stepped Reckoner
   (c) slide rule
   (d) simultaneous punching machine
   (e) Difference Engine *

13. The ______ was a more ambitious machine than the Difference Engine.
   (a) abacus
   (b) arithmetic machine
   (c) Analytical Engine *
   (d) slide rule
   (e) Stepped Reckoner

14. Which of the following is not a feature of the Analytical Engine? ______
   (a) The store
   (b) The simultaneous punching machine *
   (c) The input/output unit
   (d) System of gears and levers
   (e) The mill

15. Because of his work on the Analytical Engine ______ is considered the father of the computer.
   (a) George Scheutz
   (b) John Billings
   (c) George Boole
   (d) Charles Babbage *
   (e) Herman Hollerith

Lady Lovelace: The World's First Computer Programmer

16. ______ translated a paper on the Analytical Engine from French to English, made detailed instructions for the machine to perform certain complex calculations, and became the world's first computer programmer.
(a) Augusta Ada Byron *
(b) Thomas J. Watson, Sr.
(c) George Boole
(d) James Powers
(e) Clifford Berry

17. The first computer program was written by Augusta Ada Byron, Countess of Lovelace, while translating a paper on the ______.
(a) Stepped Reckoner
(b) Analytical Engine *
(c) simultaneous punching machine
(d) Difference Engine
(e) slide rule

Boolean Logic

18. Boolean logic is is concerned with variables that can only be ______ or ______.
(a) binary, base two
(b) digital, binary
(c) on, off
(d) zero, one
(e) true, false *

19. Binary digits are ______
(a) zero through nine
(b) true or false
(c) zero or one *
(d) numbers and letters
(e) decimal and whole numbers

Hollerith's Punched-Card Tabulating Machines

20. ______ suggested the use of punched cards for recording facts about the population that the Census Bureau must record every 10 years.
(a) Herman Hollerith
(b) John Billings *

(c) James Powers
(d) Howard Aiken
(e) Clifford Berry

21. The first patent issued for a data processing machine was given to ______.
(a) Howard Aiken
(b) John Billings
(c) James Powers
(d) Herman Hollerith *
(e) John Mauchly

22. Herman Hollerith started the Tabulating Machine Company and through mergers it became ______.
(a) Sperry Univac
(b) Remington Rand
(c) Control Data Corporation
(d) Digital Equipment Corporation
(e) International Business Machines Corporation *

Powers and the Simultaneous Punching Machine

23. ______ followed Herman Hollerith at the Census Bureau and worked on improving Hollerith's equipment while developing new machines.
(a) James Powers *
(b) John Billings
(c) Thomas J. Watson, Sr.
(d) Clifford Berry
(e) Gene Amdahl

24. The _____ allowed an operator, after typing in data, to check the data before it was entered into a card.
(a) mill
(b) input/output unit
(c) store
(d) simultaneous punching machine *
(e) punched-card control

The Dawn of the Modern Computer Age

25. Due to ______ work was speeded on the development of what is considered modern computers.
(a) World War I
(b) World War II *

(c) Spanish American War
(d) Civil War
(e) Industrial Revolution

Aiken and the Harvard Mark I

26. The dawn of the modern computer age is generally considered the completion of the ______ computer.
    (a) UNIVAC
    (b) ABC
    (c) ENIAC
    (d) Whirlwind I
    (e) Harvard Mark I *

27. The Mark I was the first successful general-purpose digital computer and the first ______ computer.
    (a) electromechanical *
    (b) stored program
    (c) timesharing
    (d) magnetic core memory
    (e) process control

Atanasoff and the ABC

28. The first computer to use the binary system for representing numbers was the ______
    (a) ENIAC
    (b) Mark I
    (c) ABC *
    (d) Whirlwind I
    (e) EDVAC

29. The father of the electronic computer is ______.
    (a) Mauchly
    (b) Aiken
    (c) Eckert
    (d) Atanasoff *
    (e) Berry

Mauchly, Eckert, and the ENIAC

30. The first electronic general-purpose computer was the ____.
    (a) ABC
    (b) ENIAC *
    (c) EDVAC
    (d) UNIVAC I
    (e) EDSAC

31. The ____ computer used decimal arithmetic.
    (a) EDSAC
    (b) ABC
    (c) Whirlwind I
    (d) EDVAC
    (e) ENIAC *

Von Neumann aand the Stored Program Concept

32. The stored program concept was devised by ____.
    (a) Mauchly
    (b) Wilkes
    (c) Von Neumann *
    (d) Berry
    (e) Aiken

33. John von Neumann joined Mauchly and Eckert on an improved version of the ENIAC called the ____.
    (a) EDVAC *
    (b) UNIVAC I
    (c) EDSAC
    (d) Whirlwind I
    (e) ABC

Wilkes and the EDSAC

34. The world's first working stored program computer was the ____.
    (a) ABC
    (b) Whirlwind I
    (c) EDVAC
    (d) EDSAC *
    (e) ENIAC

35. ____ developed the EDSAC computer.
    (a) John von Neumann
    (b) Maurice Wilkes *

(c) Jay Forrester
(d) John Mauchly
(e) Clifford Berry

UNIVAC: The First Commercial Computer

36. The first commercial data processing system was the ______ installed at the Bureau of the Census in 1951.
    (a) PDP-8
    (b) ENIAC
    (c) IBM 1401
    (d) EDSAC
    (e) UNIVAC I *

37. The ______ computer was used by CBS to forecast the presidential election of 1952.
    (a) IBM 701
    (b) CDC 6600
    (c) UNIVAC 1 *
    (d) PDP-8
    (e) EDSAC

IBM: The Giant Awakens

38. The world's most popular computer during the 1950s was the ______.
    (a) IBM 650 *
    (b) PDP-8
    (c) UNIVAC 1
    (d) Whirlwind I
    (e) IBM 701

39. IBM entered the computer field in 1953 when it introduced its model ______.
    (a) 650
    (b) 1401
    (c) 360
    (d) 701 *
    (e) 370

Generations of Computers

40. The different generations of computing have been marked primarily by major advances in ______.
    (a) terminals

(b) electronics *
(c) magnetic cores
(d) dedicated systems
(e) compatible hardware

The First Generation: 1951-1959

41. The key electrical component of first generation computers was the ______.
(a) magnetic core
(b) silicon chip
(c) vacuum tube *
(d) transistor
(e) large-scale integrated circuitry

42. ______ are tiny donut-shaped pieces of metal that can be magnetized in one of two directions.
(a) Vacuum tubes
(b) Silicon chips
(c) Integrated circuits
(d) Transistors
(e) Magnetic cores *

The Second Generation: 1959-1964

43. The ______ was the key electrical component of second generation computers.
(a) metal oxide semiconductor
(b) integrated circuit
(c) silicon chip
(d) transistor *
(e) very-large-scale integrated circuit

44. The transistor was developed by ______.
(a) Bell Laboratories *
(b) IBM
(c) Fairchild Semiconductor
(d) CDC
(e) DEC

45. The concept of ______ was designed into second generation machines allowing several users to share the machine jointly.
(a) dedicated systems
(b) multiprogramming *
(c) upwards compatibility

(d) artificial intelligence
(e) process control

46. A method of sharing a computer in which many users access a central computer using terminals is called ______.
(a) process control
(b) artificial intelligence
(c) upwards compatibility
(d) scale of integration
(e) timesharing *

47. The most significant new company to appear during the second generation was ____ which has become known for its supercomputers.
(a) IBM
(b) DEC
(c) CDC *
(d) Amdahl Corporation
(e) Cray Research Inc.

48. The ______ computer was the most successful second generation computer.
(a) IBM 1401 *
(b) PDP-11
(c) IBM 650
(d) UNIVAC 1
(e) CDC 6600

The Third Generation: 1964-1970

49. The introduction of the ______ marked the start of the third generation of computers.
(a) vacuum tube
(b) transistor
(c) minicomputer
(d) integrated circuit *
(e) metal oxide semiconductor

50. When programs that can run on a smaller model of a computer can also run on a larger model of that computer, the programs are said to be ______.
(a) dedicated
(b) upwards compatible *
(c) timesharing
(d) process control
(e) artificially intelligent

51. Integrated-circuitry made possible the production of ______.
   (a) terminals
   (b) silicon chips
   (c) minicomputers *
   (d) multiprogramming systems
   (e) metal oxide semiconductors

52. Computers that run production lines, petroleum refineries, and small businesses are called ______ computers.
   (a) Dedicated
   (b) Integrated
   (c) Magnetic core
   (d) Silicon
   (e) Process control *

The Fourth Generation: 1970-Present

53. The advent of ______ with many thousands of transistors on a silicon chip marked the start of the fourth generation of computers.
   (a) MOS
   (b) VLSI
   (c) ULSI
   (d) LSI *
   (e) PCMs

54. Core memory was replaced with ______ memory.
   (a) MOS *
   (b) PCMs
   (c) LSI
   (d) VLSI
   (e) ULSI

55. ______ build IBM-compatible hardware to compete with IBM and obtain IBM software to use with their equipment.
   (a) ULSI
   (b) LSI
   (c) PCMs *
   (d) MOS
   (e) VLSI

56. The introduction of the ______ was one of the most important developments of the fourth generation.
   (a) minicomputer
   (b) microcomputer *

(c) supercomputer
(d) process control computer
(e) digital computer

57. ______ deals with the number of transistors that can be placed on a silicon chip.
(a) Very-large-scale integration
(b) Ultra-large-scale integration
(c) Large-scale integration
(d) Scale of integration *
(e) Upwards compatibility

The Fifth Generation: 1990?

58. Breakthroughs in ______ will probably be the key to making the truly intelligent machines of the fifth generation possible.
(a) hardware
(b) memory
(c) timesharing
(d) integrated circuits
(e) software *

59. The ______ have made the realization of a fifth generation of computers by 1990 a national priority.
(a) Japanese *
(b) Americans
(c) Germans
(d) Russians
(e) French

True/False

The Abacus

1. The world's oldest known computing device is the abacus. (T p22)

2. Because of the computer revolution the abacus is no longer in use. (F p22)

Napier's Bones

3. Napier's Bones were designed to facilitate addition and subtraction. (F p22)

4. Napier's Bones led to the development of the slide rule. (T p24)

5. John Napier developed the first slide rule. (F p24)

6. The slide rule was made extinct almost overnight by the introduction of the electronic pocket calculator. (T p24)

Pascal's Arithmetic Machine

7. Pascal's arithmetic machine worked very much like an automobile odometer. (T p24)

Leibniz and the Stepped Reckoner

8. Leibniz modified Pascal's machine to perform multiplication and division directly. (T p25)

Jacquard's Loom

9. Jacquard's loom used punched-card control. (T p25)

10. Jacquard's loom operated very much like an automobile odometer. (F p25)

Babbage's Analytical Engine

11. Babbage's most ambitious calculating machine was his Difference Engine. (F p26)

12. Babbage's Difference Engine was to have a storage of 1000 numbers of 50 decimal digits each. (F p26)

13. The organization of Babbage's Analytical Engine was remarkably similar to that of modern digital computers. (T p27)

14. Babbage is remembered as the father of the computer. (T p27)

15. George Scheutz built a working Analytical Engine. (F p27)

Lady Lovelace: The World's First Computer Programmer

16. Lady Lovelace is remembered as the world's first computer programmer. (T p27)

Boolean Logic

17. Binary numbers were not adopted in the design of computer systems until the twentieth century. (T p28)

18. Geogre Boole developed his Boolean logic for the computer systems of his time. (F p28)

19. It is easier to build reliable electronics for representing binary digits than for representing decimal digits. (T p28)

Hollerith's Punched-Card Tabulating Machines

20. John Billings developed the first mechanical card-processing equipment. (F p28)

21. The first patent for a data processing machine was issued to John Billings. (F p28)

22. Thomas J. Watson founded IBM in 1896. (F p30)

Powers and the Simultaneous Punching Machine

23. James Powers's contribution was the simultaneous punching machine. (T p30)

24. James Powers founded the company that manufacturers UNIVAC computers. (T p30)

Aiken and the Harvard Mark I

25. The ENIAC was the first working realization of Babbage's Analytical Engine. (F p31)

26. The Harvard Mark I was the first successful general-purpose digital computer. (T p31)

Atanasoff and the ABC

27. The ENIAC is recognized as the first true electronic digital computer. (F p31)

28. The ABC computer was the first to use the binary system for representing numbers. (T p31)

Mauchly, Eckert, and the ENIAC

29. The ENIAC was the first electronic general-purpose computer. (T p31)

30. The ABC was a far more important machine than the ENIAC. (F p32)

Von Neumann and the Stored Program Concept

31. The stored program concept was proposed by John von Neumann. (T p32)

Wilkes and the EDSAC

32. The first working stored program computer was the EDVAC. (F p32)

UNIVAC: The First Commercial Computer

33. The first commercial computer was the UNIVAC I. (T p34)

IBM: The Giant Awakens

34. The IBM 701 was the world's most popular business computer during the 1950s. (F p35)

35. The IBM 650 was IBM's first business computer system. (F p35)

The First Generation: 1951-1959

36. The commercial computer industry began with the introduction of the UNIVAC I in 1951. (T p35)

37. Transistors were the key electronic components in first generation computers. (F p35)

38. The most successful first generation computer was the UNIVAC I. (F p35)

39. The chief reason for the IBM 650's success was magnetic core memory. (F p35)

The Second Generation: 1959-1964

40. The transistor was invented at IBM in 1948. (F p35)

41. First generation computers were dedicated systems. (T p35)

42. Control Data Corporation was the most significant new company of the second generation. (T p35)

43. The most successful second generation computer was the IBM 1401. (T p37)

The Third Generation: 1964-1970

44. The third generation was triggered by the introduction of the transistor. (F p37)

45. The IBM 1401 was a complete family of computers designed to replace all previous IBM computers. (F p37)

46. The IBM 360s were the best selling computers of the 1960s. (T p37)

47. DEC's PDP-8 was especially useful as a process control computer. (T p37)

48. The PDP-8 helped establish DEC as the leading manufacturer of minicomputers. (T p37)

49. The PDP-11 was the first computer to sell for less than $20,000. (T p38)

The Fourth Generation: 1970-Present

50. The IBM 370 was generally viewed as a revolutionary development. (F p38)

51. The companies referred to as PCM's make IBM-compatible hardware. (T p39)

52. The PDP-11 was one of the most important third generation computers. (F p39)

53. One of the most famous names in supercomputers is Seymour Cray. (T p39)

54. The minicomputer is referred to as the complete "computer on a chip." (F p39)

55. The scale of integration means how many transistors can be placed on a small silicon chip. (T p39)

The Fifth Generation: 1990?

56. The fourth generation was signaled primarily by breakthroughs in software. (F p39)

57. Germany is viewed as the primary computer competitor of the U.S. (F p39)

58. Amdahl Corporation is the industry leader in the field of supercomputers. (F p40)

Fill-in-the-Blanks

Early Computing Devices

The Abacus

1. The **abacus** is the oldest known computing device.

2. The **abacus** is a computing device that uses beads on wires to represent units, hundreds, and thousands.

Napier's Bones

3. The computing device that uses a series of sticks marked in a manner similar to today's multiplication tables is called **Napier's Bones**.

4. Using the principles of Napier's Bones, **Edmund Gunter** developed the slide rule.

5. **Napier's Bones** are marked in a manner similar to the multiplication tables of today.

Pascal's Arithmetic Machine

6. **Pascal** developed a machine that performed calculations by using cleverly connected wheels.

7. The **Pascal Arithmetic Machine** operates in the same manner as automobile odometers work as they record mileage.

Leibniz and the Stepped Reckoner

8. By modifying Pascal's machine, Gottfried Leibniz developed the **Stepped Reckoner**.

9. **Gottfried Leibniz** developed the Stepped Reckoner.

Jacquard's Loom

10. **Jacquard** invented a punched-card attachment to weaving looms that enabled the loom to repeat a design many times.

11. The first use of punched-card control was in the **Jacquard Loom**.

12. The Jacquard Loom operated under **punched-card** control.

Babbage's Analytical Engine

13. While working on the **Difference Engine**, Charles Babbage developed the idea for the Analytical Engine.

14. If it had been built, the **Analytical** Engine, would have been the world's first computer.

15. Because of his work on the Analytical Engine, **Charles Babbage** is considered the father of the computer.

16. Had the **Analytical** Engine been built, it would have been the first completely automatic, general-purpose computer.

Lady Lovelace: The World's First Computer Programmer

17. **Augusta Ada Byron, Countess of Lovelace**, is remembered as the world's first computer programmer.

18. One of the latest programming languages, **Ada**, is named after Lady Lovelace.

Boolean Logic

19. The mathematics of variables with values that can only be "true" or "false," are expressed in **Boolean** Logic.

20. It is easier to build reliable electronic circuits for representing **binary digits** than for representing decimal digits.

21. **Binary** digits consist of ones and zeros.

22. Boolean Logic helped spawn the **digital** computer.

Hollerith's Punched-Card Tabulating Machines

23. **John Billings** suggested the use of punched cards for recording facts about the population to his associate Herman Hollerith who began to design the equipment.

24. The first patent for a data processing machine was issued to **Herman Hollerith**.

25. Through a series of mergers, Herman Hollerith's Tabulating Machine Company eventually became **International Business Machines Corporation (IBM)**.

Powers and the Simultaneous Punching Machine

26. The **simultaneous punching** machine allowed an operator, after typing in data, to check and correct the data before it was actually punched into cards.

27. The **simultaneous punching** machine was developed by James Powers.

The Dawn of the Modern Computer Age

Aiken and the Harvard Mark I

28. The completion of the **Harvard Mark I** computer is generally considered to be the dawn of the modern computer age.

29. The Harvard Mark I was developed by **Howard Aiken**.

30. The first general-purpose digital computer was the **Harvard Mark I**.

31. The **Harvard Mark I** was the first electromechanical computer.

Enrichment Piece: The Mark I

32. The Mark I was controlled by **punched-paper** tape.

33. Numbers on the Mark I were stored in **mechanical switches**.

34. The Mark I used **decimal** numbers instead of binary numbers.

Atanasoff and the ABC

35. The first true electronic digital computer was the **ABC**.

36. The first machine to use the binary system for representing numbers was the **ABC**.

37. **John Vincent Atanasoff** is considered to be the father of the electronic computer.

38. The ABC stands for the **Atanasoff-Berry Computer**.

Mauchly, Eckert, and the ENIAC

39. **ENIAC** was the first electronic general-purpose computer.

40. The ENIAC was built by John W. Mauchly and **J. Presper Eckert, Jr**.

41. The ENIAC was a far more important machine than the **ABC** which often receives credit as being the first electronic digital computer.

Enrichment Piece: The ENIAC

42. When the **ENIAC** computer was turned on, it dimmed the lights of the city of Philadelphia.

43. The ENIAC used **decimal** rather than binary arithmetic.

44. A binary digit is called a **bit** for short.

Von Neumann and the Stored Program Concept

45 John von Neumann proposed the **stored program** concept.

46. The improved version of the ENIAC was called the **EDVAC**.

47. The **stored program** concept is considered among the most important developments in the computer field in the twentieth century.

Wilkes and the EDSAC

48. The world's first working stored program computer was the **EDSAC**.

49. **Maurice Wilkes** developed the EDSAC.

UNIVAC: The First Commercial Computer

50. The **UNIVAC** computer was used to forecast the election of Dwight Eisenhower as president of the United States in 1952.

IBM: The Giant Awakens

51. The first computer system introduced by IBM was the model **701**.

52. During the 1950s, the world's most popular business computer was the **IBM 650**.

Generations of Computers

53. The groups of computer systems associated with each of the significant advances in electronics have been categorized as **generations** of computers.

The First Generation: 1951-1959

54. The commercial computer industry began with the introduction of the **UNIVAC I** in 1959.

55. The key electronic component of first generation computers was the **vacuum tube**.

56. **Vacuum** tubes were bulky, consumed a great deal of electricity, gave off an enormous amount of heat, and were prone to failure.

57. Jay W. Forrester developed **magnetic core** memory.

58. The **Whirlwind I** computer was the first computer to use magnetic core memory.

59. **Magnetic cores** are tiny donut-shaped pieces of metal that can be magnetized in one of two directions.

The Second Generation: 1959-1964

60. The **transistor** replaced the vacuum tube.

61. First generation computers were **dedicated** systems, that is, only one person at a time could use the machine.

62. **Multiprogramming** allows several users to share a computer jointly.

63. Many interactive users can share a computer with the use of terminals with the concept of **timesharing**.

64. CDC is known for its enormously powerful machines that are referred to as **supercomputers**.

65. The most successful second generation computer was the **IBM 1401**.

The Third Generation: 1964-1970

66. The third generation was triggered by the development of the **integrated circuit**.

67. A family of computers in which programs that run on the smaller models can be run on the larger models is said to be **upwards compatible**.

68. The integrated circuit technology of the third generation made possible the production of powerful yet small **mini**computers.

69. Digital Equipment Corporation's PDP-8 became an extremely useful **process control** computer; thousands of PDP-8s were used to control production lines, laboratories, petroleum refineries, and even small businesses.

The Fourth Generation: 1970-Present

70. **Large-scale integration (LSI)** circuitry involves the placement of thousands of transistors on a single chip; it signaled the start of the fourth generation.

71. In fourth generation computer systems, magnetic core memory was replaced by **metal oxide semiconductor (MOS)** memories.

72. Companies that build IBM-compatible hardware to compete with IBM systems, and obtain IBM software to run on the equipment are called **plug compatible mainframers (PCMs)**.

73. The complete computer on a chip is referred to as a **microcomputer**.

74. The **scale of integration** refers to how many transistors can be placed on a small silicon chip.

The Fifth Generation: 1990?

75. It is widely believed that the fifth generation will arrive when breakthroughs in **software** make it possible to produce truly intelligent machines.

76. The **Japanese** have made the realization of the fifth generation of computers by the year 1990 a national priority.

Chapter 3
The Processor

Multiple Choice

1. ______ do not count numbers precisely.
   (a) Fixed word length computers
   (b) Digital computers
   (c) Variable word length computers
   (d) Analog computers *
   (e) Half-adders

2. ______ computers solve problems by measuring rather then by counting.
   (a) Half-adder
   (b) Fixed word length
   (c) Digital
   (d) Variable word length
   (e) Analog *

3. The brain of the computer is called the ______.
   (a) control unit
   (b) main storage
   (c) processor *
   (d) artificial intelligence
   (e) operating system

4. Another name for the processor is the ______.
   (a) CPU *
   (b) ALU
   (c) Josephson junction
   (d) auxiliary unit
   (e) ASCII

5. The field of ______ concerns machines that can be made to think.
   (a) Josephson junctions
   (b) electromagnetic pulse
   (c) biochips
   (d) artificial intelligence *
   (e) bioprocessors

The Basic Computer Processing Cycle

6. A keyboard is an example of a(n)______.
   (a) control unit
   (b) output unit
   (c) input unit *
   (d) processing unit
   (e) arithmetic and logic unit

7. A printer or video screen is considered a(n) ______.
   (a) arithmetic and logic unit
   (b) processing unit
   (c) input unit
   (d) control unit
   (e) output unit *

8. The functions a computer performs form its ______.
   (a) physical view
   (b) logical view *
   (c) computer view
   (d) arithmetic and logic unit
   (e) control unit

9. The ______ reveals how all the parts of the computer work together.
   (a) control unit
   (b) computer view
   (c) physical view *
   (d) arithmetic and logic unit
   (e) logical view

The Central Processing Unit

10. The part of the CPU that performs all mathematical calculations is the ______.
    (a) ALU *
    (b) main storage
    (c) control unit
    (d) secondary storage
    (e) instruction set

11. The part of the CPU capable of making logical decisions is the ______.
    (a) instruction set
    (b) control unit
    (c) secondary storage
    (d) ALU *
    (e) main storage

12. The ______ retains the results of the ALU's calculations until they are presented as output.
    (a) control unit
    (b) database
    (c) operating system
    (d) chip carrier
    (e) storage unit *

13. The ______ coordinates the operations of the ALU and the storage unit.
    (a) chip carrier
    (b) control unit *
    (c) operating system
    (d) secondary storage
    (e) database

14. Tapes and disks are examples of ______.
    (a) main storage
    (b) control unit
    (c) secondary storage *
    (d) central processing unit
    (e) operating system

15. Another name for secondary storage is ______ storage.
    (a) address
    (b) numeric field
    (c) character field
    (d) auxiliary *
    (e) machine

16. One of the jobs of the ______ is to make sure that the most active information is in main storage.
    (a) operating system *
    (b) control unit
    (c) ALU
    (d) CPU
    (e) database

Arithmetic and Logic Unit

17. The ______ makes decisions by comparing pieces of information to one another.
    (a) CPU
    (b) input unit
    (c) control unit
    (d) secondary storage unit
    (e) ALU *

18. A computer's built-in operations are called ______.
    (a) ALU
    (b) machine language instructions *
    (c) character set
    (d) computer programs
    (e) executions

19. The set of all of a computer's built-in operations is called a(n) ______.
    (a) operating system
    (b) numeric field
    (c) instruction set *
    (d) half-adder
    (e) character set

20. The three common comparison operations are greater than, less than, and ______.
    (a) greater than or equal to
    (b) less than or equal to
    (c) not equal to
    (d) equal to *
    (e) not less than

21. The comparison operation that determines if two values are the same is ______.
    (a) less than
    (b) equal to *
    (c) greater than
    (d) greater than or equal to
    (e) less than or equal to

22. The comparison operation that determines if one quantity is larger than another is ______.
    (a) greater than *
    (b) less than or equal to
    (c) greater than or equal to
    (d) less than
    (e) not equal to

23. The comparison operation that determines if one quantity is smaller than another is ______.
    (a) not equal to
    (b) greater than
    (c) less than or equal to
    (d) equal to
    (e) less than *

Main Storage

24. ______ are the sequences of instructions that tell the computer how to process data.
   (a) Single operand instructions
   (b) Multiple operand instructions
   (c) Computer programs *
   (d) Instruction counters
   (e) Special characters

25. A computer program must be in ______ before the computer can execute it.
   (a) auxiliary storage
   (b) ALU
   (c) secondary storage
   (d) main storage *
   (e) operating system

26. Main storage is divided into a series of ______.
   (a) addresses
   (b) values
   (c) views
   (d) fields
   (e) locations *

27. A(n) ______ is used by the computer to refer to a particular location.
   (a) address *
   (b) byte
   (c) nybble
   (d) value
   (e) field

28. The ______ is what is stored in a particular location.
   (a) field
   (b) value *
   (c) byte
   (d) address
   (e) nybble

How Microchips Are Made

29. Silicon is formed by melting ______.
   (a) plasmas
   (b) photoresist
   (c) sand *
   (d) photomask
   (e) impurities

30. Impurities that are ______ into a silicon chip determine whether a section of the chip will

conduct electricity or not.
(a) melted
(b) etched
(c) drawn
(d) engraved
(e) doped *

31. ______ is the light-sensitive material with which each silicon wafer is coated.
(a) Photomask
(b) Plasma
(c) Seed crystal
(d) Photoresist *
(e) Silicon dioxide

32. The circuits that are placed on silicon chips are designed with the help of ______ devices.
(a) CAD/CAM *
(b) EPM
(c) computer probe
(d) Josephson junction
(e) printed circuit board

33. A ______ contains the circuit pattern for one layer of the chip.
(a) photoresist
(b) silicon dioxide
(c) photomask *
(d) plasma
(e) seed crystal

34. The ionized gases that etch patterns into the silicon are called ______.
(a) photomasks
(b) plasmas *
(c) silicon dioxide
(d) seed crystals
(e) photomasks

35. Many ______ can be plugged into a printed circuit board.
(a) bytes
(b) nybbles
(c) biobits
(d) bioprocessors
(e) chip carriers *

The Binary Number System

36. Some of the early computers performed through the use of the ______ system, where the

digits zero through nine were used.
(a) binary number
(b) numeric character
(c) decimal number *
(d) flip-flop
(e) alphanumeric

37. The _____ system uses components that represent only zero or one.
(a) alphanumeric
(b) flip-flop
(c) decimal number
(d) binary number *
(e) numeric character

38. Because the computer changes a bit from a zero to a one and vice versa, a bit is also called a _____
(a) byte
(b) flip-flop *
(c) character
(d) half-adder
(e) field

Data Organization

39. The smallest unit of data people can handle conveniently is a _____.
(a) character *
(b) megabyte
(c) nybble
(d) bit
(e) field

40. The digits 0 through 9 are called _____.
(a) special characters
(b) alphanumeric characters
(c) character sets
(d) records
(e) numeric characters *

41. The letters A through Z are called _____
(a) numeric characters
(b) special characters
(c) records
(d) alphabetic characters *
(e) alphanumeric characters

42. Punctuation marks are considered ______ characters.
    (a) alhpanumeric
    (b) alphabetic
    (c) special *
    (d) numeric
    (e) register

43. EBCDIC and ASCII are examples of ______.
    (a) fields
    (b) character sets *
    (c) records
    (d) databases
    (e) files

44. EBCDIC bytes are divided into halves called ______.
    (a) nybbles *
    (b) fields
    (c) records
    (d) files
    (e) databases

45. EBCDIC bytes have two halves called the ______, and the ______.
    (a) file, record
    (b) record, field
    (c) database, file
    (d) zone, file
    (e) zone, digit *

46. A group of related characters is a ______.
    (a) zone
    (b) record
    (c) database
    (d) field *
    (e) digit

47. 1234567 is an example of a(n) ______.
    (a) alphanumeric filed
    (b) character field
    (c) numeric field *
    (d) file
    (e) record

48. A person's name that consists of only letters and blanks is an example of a(n) ______
    (a) alphabetic field *
    (b) numeric field
    (c) file

(d) character field
(e) record

49. A person's name, address, age, and height taken as a group form a(n) ______
(a) file
(b) alphanumeric field *
(c) record
(d) database
(e) alphabetic field

50. Any field that uses a special character is called a ______ field.
(a) numeric
(b) alphabetic
(c) alphanumeric
(d) character *
(e) special

51. A(n) ______ is a group of realted fields.
(a) file
(b) database
(c) record *
(d) character set
(e) address

52. A(n) ______ is a group of related records.
(a) field
(b) database
(c) address
(d) character set
(e) file *

53. A collection of related files is a(n) ______
(a) field
(b) database *
(c) address
(d) zone
(e) record

How a Computer Really Works

54. One of the most common circuits in today's computers is the ______.
(a) zone
(b) nybble
(c) half-adder *
(d) register

(e) branch

55. The three fundamental circuit components are AND, OR, and ______.
(a) NOT *
(b) BIPS
(c) LESS
(d) ELSE
(e) EQUAL

Main Storage Size and Addressing

56. Main storage is measured in units of ______ bytes.
(a) X
(b) P
(c) BIPS
(d) K *
(e) KKK

57. K equals ______.
(a) 512
(b) 1024 *
(c) 4096
(d) 1024 x 1024
(e) 16 x 1024

58. Mb stands for ______.
(a) minibytes
(b) maxibytes
(c) megabytes *
(d) memorybytes
(e) memorybits

59. Instructions that manipulate information in main storage must include tha address of the start of the data field, the length of the data field in bytes, and the ______ to be performed.
(a) zone
(b) record
(c) database
(d) file
(e) operation *

Editing

60. The process in which data is prettied up for presentation is called ______.
(a) nondestructive read-out

(b) editing *
(c) looping
(d) accumulating
(e) destructive read-in

Destructive Read-In and Nondestructive Read-out

61. ______ occurs when the processor places data in main storage without looking to check if there is information in that location.
(a) destructive read-in *
(b) looping
(c) editing
(d) destructive read-out
(e) nondestructive read-out

62. The process of reading data out of a storage location is called ______.
(a) nondestructive read-in
(b) destructive read-in
(c) editing
(d) nondestructive read-out *
(e) looping

Machine Language Instructions

63. The various operations a computer can interpret and perform are generally referred to as ______.
(a) single operand instructions
(b) registers
(c) machine language instructions *
(d) multiple operand instructions
(e) accumulators

64. Computer instructions that reference only one piece of data are called ______.
(a) accumulators
(b) machine language instructions
(c) registers
(d) multiple operand instructions
(e) single operand instructions *

65. Computer instructions that reference two or more data items are called ______.
(a) multiple operand instructions *
(b) registers
(c) machine language instructions
(d) accumulators

(d) single operand instructions

66. A temporary holding place in the ALU for instructions is called a ______.
    (a) variable
    (b) register *
    (c) branch
    (d) jump
    (e) constant

67. Another name for a register is ______
    (a) constant
    (b) jump
    (c) accumulator *
    (d) variable
    (e) branch

The Instruction Execution Cycle

68. The ______ contains the location of the next instruction to be executed.
    (a) constant
    (b) jump
    (c) accumulator
    (d) variable
    (e) instruction counter *

69. The computer fetches an instruction and places it in the ______.
    (a) instruction counter
    (b) variable
    (c) fetch
    (d) instruction register *
    (e) machine language instruction

70. When an overflow occurs we have what is called a(n) ______.
    (a) loop
    (b) fatal error *
    (c) accumulator
    (d) register
    (e) branch

Variable Word-Length and Fixed Word-Length Machines

71. A computer that allows a field to occupy as many bytes as needed is called a ______ machine
    (a) single operand
    (b) fixed word-length

(c) multiple operand
(d) looping
(e) variable word-length *

72. ______ machines perform their operations in terms of words rather than individual bytes.
(a) Single operand
(b) Variable word-length
(c) Fixed word-length *
(d) Looping
(e) Multiple operand

Machine Lnaguage Programming

73. Computers normally perform instructions sequentially but they can jump or ______ to another instruction.
(a) branch *
(b) read
(c) execute
(d) load
(e) run

74. A location whose value changes is called a ______.
(a) constant
(b) branch
(c) loop
(d) variable *
(e) comparisin

75. A location whose value remains fixed is called a ______.
(a) variable
(b) register
(c) constant *
(d) loop
(e) branch

Looping: The Real Power of the Computer

76. ______ allows the computer to reuse certain instructions.
(a) Branching
(b) Looping *
(c) Loading
(d) Jumping
(e) Adding

77. By using a ______, instructions can be used many times in a program.
    (a) comparison
    (b) branch
    (c) jump
    (d) load
    (e) loop *

The Josephson Junction: The "Computer on Ice"

78. ______ is a data processing switch that is designed to operate at extremely low temperatures.
    (a) Biochip
    (b) EMP
    (c) Josephson junction *
    (d) Bioprocessor
    (e) BIPS

Consequences of Technology

Biochips (Fleshware?)

79. Computer scientists, genetic engineers, and microelectronic engineers are developing ______ that are made from proteins.
    (a) plasmas
    (b) silicon chips
    (c) photoresist
    (d) biochips *
    (e) Josephson junctions

80. A new type of computer chip made from proteins is called a ______.
    (a) biobit
    (b) biochip *
    (c) biocpu
    (d) Josephson junction
    (e) biobrain

True/False

Introduction

1. Digital computers perform calculations by counting numbers precisely. (T p48)

2. The abacus is a digital device. (T p48)

3. Analog computers solve problems by measuring. (T p48)

4. Analog computers measure precisely, but digital computers can only count to a certain level of accuracy. (F p48)

5. It is reasonable to say today that the world is "going analog." (F p48)

6. The processor is the "brain" of the computer. (T p48)

The Basic Computer Processing Cycle

7. The keyboard of a computer terminal is an input unit. (T p48)

8. Printers and video screens are output units. (T p48)

9. A logical view of the computer reveals how the computer's actual "knobs, buttons, bells, and whistles" all work together. (F p49)

The Central Processing Unit

10. The three logical units of the CPU are the arithmetic and logic unit (ALU), main storage, and the control unit. (T p49)

11. The ALU performs arithmetic operations and makes logical decisions. (T p49)

12. Most information is normally kept in main storage. (F p49)

13. The most common main storage devices are tapes and disks. (F p49)

14. When information is to be processed, the computer transfers the information from main storage to secondary storage. (F p49)

15. The operating system makes sure that active information is kept in main storage. (T p49)

Arithmetic and Logic Unit

16. A computer's built-in operations are called machine language instructions. (T p51)

17. The three common comparison operations are equal to, not equal to, and less than. (F p51)

Main Storage

18. A computer program must be in auxiliary storage for the computer to execute it. (F p51)

19. Data must be in main storage to be processed by an executing program. (T p51)

20. Every main storage location has an address and a value. (T p51)

Enrichment Piece: How Microchips Are Made

21. Microchips are made from rare materials. (F p52)

22. Impurities are actually critical to the operation of microchips. (T p52)

23. Silicon wafers are coated with a special light-sensitive material called photomask. (F p52)

24. Some systems etch the circuit patterns onto silicon chips by using ionized gases called plasmas. (T p53)

25. It's cheaper to throw away flawed chips than to fix them. (T p54)

26. A silicon chip will crumble like a cookie if dropped. (T p54)

27. Most of the U.S. firms that make microchips are located in an area of Texas called Silicon Valley. (F p56)

The Binary Number System

28. Bits are often called flip-flops. (T p58)

29. Today's fastest computers process trillions of operations per second. (F p58)

Data Organization

30. The smallest unit of data that people normally handle is a character. (T p58)

31. Blanks are generally treated as numeric characters. (F p58)

32. Punctuation marks are examples of special characters. (T p58)

33. With three bits, eight bit-patterns are possible. (T p60)

34. Six bits can represent 128 characters. (F p60)

35. The two character sets that have become widely standardized are EBCDIC and ASCII. (T p60)

36. In EBCDIC, half a byte is called a nybble. (T p60)

37. In EBCDIC, the rightmost four bits are the zone portion of the byte. (F p60)

38. A group of related characters is called a field. (T p60)

39. An alphabetic field contains only letters of the alphabet. (F p60)

40. An alphanumeric field contains only letters and digits. (F p60)

41. A file is a group of related records. (T p61)

Enrichment Piece: How a Computer Really Works

42. The three fundamental circuit operations are AND, OR, and NOT. (T p59)

43. The binary equivalent of 2 is 11. (F p59)

44. The following describes the AND circuit operation: If both inputs are 0, then the output is 0; otherwise the output is 1. (F p59)

Main Storage Size and Addressing

45. In computer applications, K equals 1000. (F p61)

46. In computer applications, the letter M equals 1024 times 1024. (T p61)

47. A 4K computer would have addresses ranging from 0 to 4096. (F p61)

Editing

48. "Prettying up" 412.1 to the dollar amount format $412.10 is called editing. (T p62)

Destructive Read-In and Nondestructive Read-Out

49. The process of storing information in main storage is called nondestructive read-out. (F p62)

50. When the processor reads information from a storage location, it makes a copy of the data while leaving the original data intact. (T p62)

Machine Language Instructions

51. Instructions that reference two or more data items are called multiple operand instructions. (T p62)

52. Most programming today is done in machine language. (F p63)

The Instruction Execution Cycle

53. The accumulator always contains the location address of the next instruction to be executed. (F p63)

54. Most computers will terminate a program when an overflow error occurs. (T p64)

55. Overflow errors are generally considered to be fatal errors. (T p64)

Variable Word-Length and Fixed Word-Length Machines

56. Fixed word-length machines perform their operations in terms of words rather than individual bytes. (T p64)

57. Variable word-length machines are more convenient for processing text, where words of different lengths are processed. (T p64)

Machine Language Programming

58. Looping greatly reduces the number of instructions the programmer must write. (T p66)

The Josephson Junction: The "Computer on Ice"

59. Josephson Junction computers operate on the principle of superconductivity. (T p67)

Biochips (Fleshware?)

60. One problem with biochips is that they are extremely susceptible to electromagnetic pulse. (F p68)

Fill-in-the-Blanks

1. **Digital** computers perform calculations by counting numbers precisely.

2. **Analog** computers solve problems by measuring.

3. Every digital computer contains a **central processing unit (CPU)**, or processor.

4. The **CPU** of a digital computer is the hardware that guides the computer through the various steps it takes to solve problems.

5. The processor is also called the **brain** of the computer.

6. The field of **artificial intelligence** is developing machines that exhibit what appears to be intelligent behavior.

The Basic Computer Processing Cycle

7. A keyboard is an example of an **input** unit.

8. Information is entered into a computer through an **input** unit.

9. Results from the computer are displayed via an **output** unit.

10. A video display screen is an example of an **output** unit.

11. The functions that a computer performs comprise its **logical** view.

12. A **physical** view of the computer reveals how the computer's actual knobs, buttons, bells, and whistles all work together.

The Central Processing Unit

13. The **arithmetic and logic unit (ALU)** performs such operations as addition, subtraction, multiplication, and division.

14. Logical decisions are made by the **arithmetic and logic unit (ALU)** of the computer.

15. **Main storage** stores information that arrives via the input unit so that this information is available to the ALU when calculations are to be performed.

16. The **control unit** coordinates the operations of the ALU and main storage.

17. Active information is retained in **main storage**.

18. Main storage is relatively expensive, so most information is held in **secondary or auxiliary** storage.

19. Another name for secondary storage is **auxiliary storage**.

20. When information is to be processed, the computer transfers the information from secondary storage to **main storage**.

21. One of the jobs of the **operating system** of a computer is to make sure that the most active information is in main storage.

Arithmetic and Logic Unit

22. The **ALU** can actually make decisions; it does so by comparing pieces of information to one another.

23. A computer's built-in operations are called **machine language instructions**, instructions, or machine instructions.

24. The set of all a computer's built-in operations is called its **instruction set**.

25. The comparison operation that determines if two pieces of information are identical is the **equal to** operation.

26. The computer determines if one quantity is larger than another by using the **greater than** comparison operation.

27. Using the **less than** comparison operation a computer can determine if one quantity is smaller than another.

28. By combining the equal to, greater than, and less than comparison operations, three other popular comparison operations can be formed; these are **not equal to**, less than or equal to, and greater than or equal to.

Main Storage

29. **Computer programs** are sequences of instructions that tell the computer what to do.

30. For the computer to be able to execute a computer program, the program must first be in **main storage**.

31. Data must be in **main storage** in order to be processed by an executing program.

32. Main storage is divided into a series of **locations**.

33. A **location** in main storage contains either a data item or an instruction.

34. Every location has an **address** and a value.

35. The **value** is what is stored in a location.

Enrichment Piece: How Microchips Are Made

36. At the heart of today's computer systems is the **microchip**.

37. Microchips are made of **silicon** that comes from sand heated to an extremely high temperature.

38. Conducting and nonconducting regions make an area of a chip behave just like a **transistor**.

39. **Photoresist** is a light-sensitive material used to coat silicon wafers.

40. The circuits that are placed on a silicon chip are designed by computer with the help of **CAD/CAM** equipment.

41. A **photomask** contains the circuit pattern for one layer of a chip.

42. Circuit designs can be etched onto a silicon chip using ionized gases called **plasmas**.

43. For protection, each chip is bonded to a **chip carrier**.

44. **Circuit boards** are made up of several hundred separate chips.

45. Most chips made in the United States are produced in an area outside San Francisco called **Silicon Valley**.

The Binary Number System

46. Early computers used the **decimal** number system, with electrical components representing digits from zero through nine.

47. Computer using the **binary number** system, have electrical components that represent only zeros or ones.

48. A bit can change from a "1" to a "0" or vice versa, so bits are often called **flip-flops**.

Data Organization

49. The smallest unit of data that people normally handle is a **character**.

50. A character is also called a **byte**.

51. **Numeric** characters are the digits 0 through 9.

52. **Alphabetic** characters are the letters of the alphabet in either uppercase or lowercase, and the blank.

53. A blank is considered an **alphabetic** character.

54. Characters such as a period, question mark, percent sign, and dollar sign are called **special** characters.

55. A **character set** is a list of all the characters that can be used with a particular computer, as well as the bit patterns that correspond to each character.

56. The two most popular character sets are EBCDIC and **ASCII**.

57. The **EBCDIC** character set uses patterns of eight bits per character.

58. The **ASCII** character set is used on most personal computers and for data communications.

59. In the EBCDIC character set, each character or byte is divided into two halves called **nybbles**.

60. The leftmost four bits of an EBCDIC character are called the **zone** portion.

61. In the EBCDIC character set, the rightmost four bits are called the **digit** portion.

62. A group of related characters is called a **field**.

63. A **numeric** field contains only digits.

64. A person's name is an example of an **alphabetic** field.

65. An **alphanumeric** field would consist of letters, spaces, and digits.

66. **$**42,315.82 is an example of a **character** field.

67. A group of related fields is called a **record**.

68. A **file** is a group of related records.

69. A **database** may be viewed as a collection of related files.

Enrichment Piece: How a Computer Really Works

70. Computer circuits are built from three fundamental components; **AND**, OR, and NOT.

71. One of the most common circuits used in today's computers is the **half-adder**.

72. The **half-adder** computer circuit adds two binary digits producing a sum and a carry, if necessary.

73. If both inputs into an AND are 1, then the output is **1**, otherwise the output is 0.

74. If the input into a NOT is 0, the output is **1**.

Main Storage Size and Addressing

75. The larger the main storage size, the **more**/less powerful the computer.

76. **Main storage** is measured in units of K bytes.

77. In computer applications K equals **1024**.

78. **Megabyte** can be written as Mb.

79. A **4K** computer would have addresses ranging from 0 to 4095.

80. Instructions that manipulate information in the computer's main storage must include the address of the start of the data field, the **length of the data field** in bytes, and the operation to be performed upon the data.

Editing

81. The reorganization, rearrangement, or "pretting up" of computer results is called **editing**.

Destructive Read-In and Nondestructive Read-Out

82. **Destructive read-in** occurs when the processor stores data in a main storage location that already contains information.

83. The process or reading data out of a storage location is often called **nondestructive read-out** .

84. Making a new recording over an existing one is an example of **destructive read-in** .

85. An example of **nondestructive read-out** is the replaying of a recording many times.

Machine Language Instructions

86. **Single operand** instructions reference only one piece of data.

87. With **multiple operand** instructions, computers can specify two or more data items to be operated upon.

88. The various operations a computer can interpret and perform are called its **machine language** instructions.

89. A load instruction places data into a register in the ALU called the **accumulator** .

The Instruction Execution Cycle

90. The **instruction counter** contains the location in main storage of the next instruction to be executed.

91. The **instruction** register contains the instruction being performed.

92. The computer **fetches** an instruction from main storage and places it in the instruction register.

93. If a multiplication result is larger than the storage location it is assigned to, an **overflow** error occurs.

94. Most computers will terminate a program when a serious error occurs; these erors are called **fatal** errors.

95. The **CPU** of a computer may be likened to the operating room of a hospital.

Variable Word-Length and Fixed Word-Length Machines

96. **Variable word-length** machines allow fields to occupy as many bytes as needed, within certain limits.

97. **Fixed word-length** machines process all information as fixed-sized groups of bytes.

98. **Fixed word-length** machines are less flexible, but they can perform operations faster.

Machine Language Programming

99. Computers are **general** purpose machines that are capable of performing instructions supplied by computer programs.

100. Normally, instructions are performed **sequentially**, that is, in order.

101. A computer can be instructed to perform an instruction located in a different part of a program; this allows the computer to jump or **branch** to the other instruction.

102. When the value of a location can change as a program runs, that location is called a **variable**.

103. A location that contains a fixed value is called a **constant**.

Looping: The Real Power of the Computer

104. With **looping** a computer can reuse certain instructions many times.

105. **Looping** greatly reduces the number of instructions a programmer has to write.

The Josephson Junction: The "Computer on Ice"

106. The **Josephson junction** is a data processing switch designed to operate at extremely low temperatures.

Biochips (Fleshware?)

107. The **biochip** is a computer chip made from proteins.

108. Computer scientists, genetic engineers, and microelectronic engineers are working to create **biochips**, or computers made from proteins.

109. **Electromagnetic pulse (EMP)** would cause electronic instrumentation to malfunction after a nuclear explosion.

Chapter 4
Input: Gateway to the Computer

Multiple Choice

1. Reading data into the computer is called ______.
   (a) GIGO
   (b) output
   (c) CAI
   (d) input *
   (e) processing

2. ______ devices are built separately from the processing unit and main storage of the computer.
   (a) Peripheral *
   (b) Control
   (c) Turnaround
   (d) Arithmetic and logic
   (e) Cursor

Punched Cards

3. The phrase, "Do not bend, fold, spindle, or mutilate," refers to ______.
   (a) simultaneous punching machine
   (b) punched cards *
   (c) keypunching
   (d) card verification
   (d) card readers

4. Which of the following is not an advantage of the 96-column card? ______
   (a) Less costly to store
   (b) Easier to handle
   (c) Smaller than the 80-column card
   (d) Cheaper than the 80-column card
   (e) Has file integrity *

Garbage-In-Garbage-Out

5. The term ______ refers to the fact that if the improper information is placed in the

computer, the information the computer delivers is likely to be incorrect.
(a) CAI
(b) MICR
(c) GIGO *
(d) UPC
(e) OMR

The Keypunch Machine

6. Information punched into cards by a keypunch machine comes from a ______
(a) source document *
(b) page reader
(c) OMR
(d) digitizer
(e) menu

7. One card at a time is moved to the ______ where the operator types the data to appear on the card.
(a) card verifier
(b) digitizer
(c) menu
(d) punching station *
(e) input hopper

8. The ______ machine has a typewriter like keyboard and hole punching mechanism.
(a) timesharing
(b) key-to-disk
(c) keypunch *
(d) key-to-tape
(e) card verifier

Card Verification

9. A ______ is used to ensure that keypunched data is correct.
(a) keypunch machine
(b) card verifier *
(c) punching station
(d) output hopper
(e) input hopper

10. The operator of the ______ device should not be the same person who did the keypunching.
(a) key-to-disk
(b) direct data entry
(c) key-to-tape

(d) punching station
(e) card verifier *

Eliminating Verification

11. ______ checks ensure that digits are not entered where letters are required.
(a) Reasonableness
(b) Range
(c) Card
(d) Edit *
(e) Optical

12. A check to determine if the correct month number has been entered is called a ______ check.
(a) range *
(b) edit
(c) optical
(d) reasonableness
(e) card

13. Cards that fail a ______ check are not always incorrect.
(a) optical
(b) edit
(c) reasonableness *
(d) range
(e) card

Reading Cards into the Computer

14. Once cards have been punched and verified a device called a ______ enters the information from the cards into the computer.
(a) read station
(b) input hopper
(c) CRT
(d) source document
(e) card reader *

15. The card reader has an input hopper, output hopper and a ______.
(a) CRT
(b) read station *
(c) wand reader
(d) page reader
(e) tape reader

An Assessment

16. The primary data entry medium through the early 1970s was ______.
    (a) key-to-disk
    (b) magnetic ink characters
    (c) key-to-tape
    (d) punched cards *
    (e) paper tape

17. Because it is easy to remove cards from their files ______ problems can develop.
    (a) range check
    (b) edit check
    (c) file integrity *
    (d) reasonableness
    (e) even parity

Key-to-Tape and Key-to-Disk Systems

18. Key-to-tape and key-to-disk machines are called ______ because they have their own built-in computers.
    (a) intelligent data entry machines *
    (b) terminal-oriented systems
    (c) transaction-processing systems
    (d) membrane keyboards
    (e) multifont readers

Terminal-Oriented Systems

19. An example of operating on-line with a computer is ______.
    (a) keypunching
    (b) key verification
    (c) batch processing
    (d) turnaround document processing
    (e) timesharing *

20. An example of working off-line from a computer is ______ processing.
    (a) digital
    (b) batch *
    (c) page
    (d) wand
    (e) document

21. On-line, terminal-oriented systems are often called ______ systems.
    (a) key-to-disk

(b) key-to-tape
(c) batch processing
(d) transaction-processing *
(e) touch sensing

Touch Sensing

22. With ______ the user can point to an item on a computer screen.
(a) transaction processing
(b) wand readers
(c) touch sensing *
(d) digital data
(e) membrane keyboards

23. When using touch sensing a ______ menu is often used to list the topics available to the user.
(a) reversed field
(b) joystick
(c) light pen
(d) cursor
(e) menu *

24. Four ways to enter information from a touch screen are touching the screen directly, using a light pen, using a joystick and using a ______.
(a) data tablet *
(b) journal tape reader
(c) input hopper
(d) cursor
(e) bar code reader

25. CRTs are examples of ______ devices.
(a) speech recognition
(b) soft copy *
(c) hard copy
(d) input
(e) batch processing

26. In order to get a hard copy from a CRT, the CRT is normally connected to a ______.
(a) terminal
(b) read station
(c) output hopper
(d) printer *
(e) page reader

Other Input Devices

27. A ______ device can convert verbal commands into computer-stored data.
   (a) membrane keyboard
   (b) speech synthesis
   (c) speech recognition *
   (d) OCR
   (e) MICR

28. ______ can be used to enter pictures into a computer.
   (a) MICR
   (b) OCR
   (c) Membrane keyboards
   (d) OMR
   (e) TV cameras *

29. The ______ input device is a flat surface that displays pictures and symbols.
   (a) membrane keyboard *
   (b) light pen
   (c) MICR
   (d) OCR
   (e) OMR

Source Data Automation

30. Which of the following is not a type of SDA? ______
   (a) OCR
   (b) OMR
   (c) CAI *
   (d) MICR
   (e) Bar code reading

31. The ______ is the SDA technique that automated check processing.
   (a) bar code reading
   (b) MICR *
   (c) OCR
   (d) OMR
   (e) CAI

32. The SDA technique that allows a computer to read human-readable symbols is ______.
   (a) CAI
   (b) bar code reading
   (c) OMR
   (d) OCR *
   (e) MICR

33. ______ is the SDA technique commonly used in exam scoring applications.
   (a) OCR
   (b) CAI
   (c) Bar code reading
   (d) MICR
   (e) OMR *

34. The ______ SDA technique is commonly used in inventory and checkout systems.
   (a) bar code reading *
   (b) OMR
   (c) CAI
   (d) OCR
   (d) MICR

35. With ______ is captured at its source and entered into the computer without any paper documents being produced.
   (a) file integrity
   (b) computer-assisted instruction
   (c) transaction processing
   (d) direct data entry *
   (e) journal tape

Magnetic Ink Character Recognition

36. ______ was developed by the banking industry.
   (a) OCR
   (b) MICR *
   (c) CAI
   (d) OMR
   (e) Bar code reading

37. MICR readers are sensitive to ______.
   (a) even parity
   (b) dark stripes
   (c) magnetized ink *
   (d) digital data
   (e) holes punched in cards

Optical Character Recognition

38. ______ devices can read hand-written or typewritten characters.
   (a) Bar code readers
   (b) OMR

(c) MICR
(d) OCR *
(e) CAI

39. Printed-font readers that are used to read credit card impressions are generally ______.
(a) journal tape readers
(b) multifont readers
(c) page readers
(d) wand readers
(e) single-font readers *

40. ______ readers are used in data processing operations that service many different types of OCR applications.
(a) Page
(b) Single-font
(c) Wand
(d) Multifont *
(e) Journal tape

41. The most versatile and most expensive printed-font reader is the ______.
(a) omnifont reader *
(b) journal tape reader
(c) single-font reader
(d) multifont reader
(e) page reader

42. The Kurzweil OCR reader is an example of a ______ reader.
(a) wand
(b) multifont
(c) omnifont *
(d) single-font
(e) journal-tape

43. ______ readers are used in the retail trades and are capable of reading cash register printouts.
(a) Wand
(b) Journal-tape *
(c) Single-font
(d) Omnifont
(e) Multifont

44. The ______ reader is used to read utility bills.
(a) wand
(b) page
(c) journal-tape
(d) bar code

(e) small document *

45. ______ readers can read full sheets of text that have a variety of fonts.
(a) Bar code
(b) OMR
(c) Page *
(d) Wand
(e) MICR

46. The ______ reader is a hand-held unit that brings OCR capabilities to the source of the data.
(a) wand *
(b) page
(c) bar code
(d) journal-tape
(e) small document

How OCR Works

47. Light is reflected onto lightsensitive ______ that register the presence or absence of light in an OCR reader.
(a) reversed fields
(b) digitzers
(c) screens
(d) lasers
(e) photocells *

48. ______ devices should not be used in situations where extremely high accuracy is needed.
(a) OMR
(b) OCR *
(c) MICR
(d) UPC
(e) CAI

The Kurzweil OCR Reader

49. The Kurzweil OCR reader has a built-in ______ that allows it to check words with unrecognizable characters.
(a) laser
(b) digitizer
(c) dictionary *
(d) menu
(e) membrane

50. A version of the Kurzweil OCR reader that incorporates ______ is being used by the blind.

(a) speech recognition
(b) laser holography
(c) speech synthesis *
(d) three-dimensional digitation
(e) timesharing

OCR Applications

51. OCR readable merchandise tags are normally read by ______ at the time of a purchase.
(a) single-font readers
(b) wand readers *
(c) page readers
(d) multifont readers
(e) omnifont readers

52. A ______ document is a computer-prepared document that have additional information added and then be resubmitted to the computer.
(a) turnaround *
(b) source
(c) hard copy
(d) soft copy
(e) parity

Optical Mark Recognition

53. The form of OCR that is used to record answers to multiple choice exams is ______.
(a) SDA
(b) MICR
(c) CAI
(d) OCR-A
(e) OMR *

54. By combining OMR with ______ bills can be processed that contain both computer printed information and pencil marked data.
(a) MICR
(b) CAI
(c) holography
(d) OCR *
(e) UPC

Bar Code Reading

55. One of the fastest growing areas in source data automation for inventory applications is

_____.
(a) laser holography
(b) optical mark recognition
(c) bar code reading *
(d) magnetic ink recognition
(e) optical character recognition

Universal Product Code

56. Most items sold in supermarkets have _____ on their labels.
(a) reversed parity
(b) universal product code *
(c) magnetic ink characters
(d) digital data
(e) reversed fields

57. The UPC code uses a 10 digit scheme; the five leftmost digits identify the _____.
(a) item
(b) price
(c) weight
(d) size
(e) manufacturer *

58. The five rightmost digits on the UPC code identify the _____.
(a) price
(b) manufacturer
(c) item *
(d) size
(e) color

59. The _____ scheme used in UPC coding helps to determine if the machine has read the code properly.
(a) even parity *
(b) laser holography
(c) item pricing
(d) reversed field
(e) OCR

Benefits of UPC Scanning

60. UPC scanning can monitor sales of items to prevent _____.
(a) slow checkouts
(b) stockouts *
(c) item pricing

(d) price hikes
(e) comparison shopping

61. Data from UPC scanning can help stores schedule ______.
    (a) stockouts
    (b) item pricing
    (c) comparison shopping
    (d) wages
    (e) employee hours *

Controversial Issues in UPC Scanning

62. The most controversial issue surrounding UPC scanning is ______.
    (a) stockouts
    (b) employee hours
    (c) comparison shopping
    (d) item pricing *
    (e) wages

63. Which of the following is not a key concern of advocates of item pricing? ______
    (a) Meal planning
    (b) Comparison shopping
    (c) Stockouts *
    (d) Budgeting
    (e) Price verification

Directions in UPC Scanning

64. The use of ______ will enable UPC scanners to read the UPC codes off items no matter how they are positioned over the scanner.
    (a) laser holography *
    (b) even parity
    (c) wand readers
    (d) item pricing
    (e) digitizers

True/False

Introduction

1. Input, output, and secondary storage devices are called peripheral devices. (T p75)

Punched Cards

2. The most common card format is the 96-column card with circular holes. (F 75)

Card Verification

3. The card verifier device actually repunches the holes a second time to be sure they are correct. (F p77)

4. Although card verification is expensive, it is worthwhile because it ensures 100 percent accuracy. (F p77)

5. The card verification operator should not be the same person as the keypunch operator. (T p77)

Eliminating Verfication

6. Edit checks ensure that digits are not entered where letters are required and vice versa. (T p77)

7. Range checks ensure only that a data item is reasonable. (F p77)

8. An example of an edit check is checking that a person's age is "less than 150." (F p77)

An Assessment

9. The use of punched cards has been declining. (T p77)

10. Because individual cards can be lost or stolen, and extra cards can be added, cards are said to have file integrity problems. (T p78)

Key-to-Tape and Key-to-Disk Systems

11. Card punch machines and card verification machines are generally considered to be examples of intelligent data entry devices. (F p78)

Terminal-Oriented Systems

12. In timesharing systems, users operate on-line to the computer. (T p78)

13. Batch processing is considered a means of off-line operation. (T p79)

Touch Sensing

14. Touch sensing is generally considered to be a user-friendly means of input. (T p79)

15. Ideally, about 30 options should be displayed on a touch-sensing screen at once. (F p79)

16. CRTs are called soft copy devices. (T p82)

Other Input Devices

17. A TV camera can be used as a computer input device. (T p82)

18. A membrane keyboard has no movable keys. (T p82)

19. No input devices can as yet input three-dimensional information, but such devices will be common in a decade. (F p82)

Source Data Automation

20. Bypassing automated data entry is called source data entry. (F p84)

21. OCR is used primarily in automated check processing. (F p84)

22. MICR is used in applications where the computer reads human-produced symbols directly. (F p84)

23. OMR is commonly used in exam-scoring applicatioms. (T p84)

24. Bar code reading is commonly used in supermarket-checkout systems. (T p85)

25. With direct data entry, no paper documents are produced. (T p85)

Magnetic Ink Character Recognition

26. MICR readers generally process 50,000 to 100,000 checks per minute. (F p85)

27. The "on us" symbol is a common MICR symbol. (T p85)

Optical Character Recognition

28. Multifont OCR readers are the most versatile and most expensive OCR devices on the market today. (F p86)

29. Omnifont OCR readers are often equipped with complex programs that help them learn new fonts. (T p86)

30. The most widely used OCR font in the United States is OCR-B. (F p86)

31. Most people find OCR-A easier to read than OCR-B. (F p86)

32. Studies have shown that OCR-A is recognized by machines far more reliably than OCR-B. (T p86)

How OCR Works

33. OCR reading of hand printing should not be used in situations where very high accuracy is critical. (T p88)

OCR Applications

34. A turnaround document, when read back into the computer, contains only computer-printed data. (F p90)

35. Turnaround documents are especially effective in inventory control applications. (T p90)

The Kurzweil OCR Reader

36. The Kurzweil OCR reader executes approximately 80,000 instructions to recognize a single character. (T p89)

37. The Kurzweil OCR reader requires that the operator identify every unreadable character. (F p89)

38. The Kurzweil OCR reader can "learn" new characters for future reference. (T p89)

Optical Mark Recognition

39. With OMR, pencil marks are recognized in particular positions on specially designed forms. (T p90)

Combining OCR and OMR

40. In computerized billing applications combining OCR and OMR, the amount paid by the customer is always entered manually and in the OMR portion of the bill. (F p91)

Bar Code Reading

41. It is important that scanning of bar coded information be done at an even pace. (T p91)

Universal Product Code

42. The UPC is a bar code scheme consisting of 12 digits. (F p94)

43. The leftmost digits in a UPC item code identify the manufacturer of the item. (T p94)

44. The UPC coding scheme uses odd parity, that is, all representations of the digits have an odd number of ones (or bars). (F p94)

45. Today's UPC scanners make about one error per several thousand digits. (F p94)

46. When an item sells "two for 79 cents" the UPC scanner charges 39 cents for the first one it sees and 40 cents for the second. (F p94)

Controversial Issues in UPC Scanning

47. Item pricing is perhaps the most controversial issue surrounding UPC scanning. (T p96)

48. Supermarkets favor item pricing, but consumer groups wish to eliminate it. (F p96)

Directions in UPC Scanning

49. UPC scanners using laser holography can read UPC codes regardless of how a package is positioned. (T p96)

50. The item pricing controversy may eventually be solved by economical devices that read the UPC codes and automatically stamp the prices on packages. (T p96)

Fill-in-the-Blanks

1. Getting data into the computer is called **input**.

2. Devices that are built separately from the processing unit and the main storage of the

computer are called **peripheral** devices.

Punched Cards

3. The most common punched card format has **80** columns.

4. With the System/3, IBM introduced the smaller **96** column card.

Enrichment Piece: Garbage-In-Garbage-Out

5. **GIGO** is the computer term that indicates what the result is likely to be when incorrect information is entered into the computer.

The Keypunch Machine

6. Data from a **source** document is punched onto cards by a human operator using a keypunch machine.

7. The machine that punches holes into cards is called a **keypunch** machine.

8. Cards are punched with holes at the **punching** station of a keypunch machine.

Card Verification

9. The machine that checks to ensure that information punched onto cards is correct is a **card verifier**.

10. In card **verification** the operator should not be the same person as the keypunch operator.

Eliminating Verification

11. **Edit** checks ensure that digits are not entered where letters are required.

12. **Range** checks ensure that a data item has a value between specific values.

13. Data that fails a **reasonableness** check is not necessarily incorrect.

Reading Cards into the Computer

14. Punched cards are entered into the main computer system through a device called a **card reader**.

15. Punched cards are read one at a time as they pass through the **read** station of a card reader.

16. After a card has been read by the card reader it is placed in an **output** hopper.

17. Cards waiting to be read by the computer are placed in an **input** hopper.

An Assessment

18. The primary data entry medium through the early 1970s was **punched cards**.

19. Because it is easy to misplace or remove a card from a card file, cards are said to have **file integrity** problems.

Key-to-Tape and Key-to-Disk Systems

20. Key-to-tape and key-to-disk machines are sometimes called **intelligent data entry** devices because they have built-in computers.

21. **Key-to-tape and key-to-disk** machines enter computerized data directly onto reusable magnetic tapes and disks.

Terminal-Oriented Systems

22. Terminals or keyboards that are directly connected to a computer or data communications network are said to be **on-line** to the computer or network.

23. **Timesharing** allows hundreds and even thousands of users to be on-line with a single computer system.

24. With punched cards, users operate **off-line** from their computer systems.

25. Users who submit their work and receive results through devices not directly connected to the computer, use a method that is called **batch** processing.

26. In **batch processing** each group of information is called a batch.

27. On-line, terminal-oriented timesharing systems are often called **transaction-processing** systems.

28. In a **transaction-processing** system, each data item is called a transaction.

Touch Sensing

29. Operators can enter information into a computer by pointing to an item displayed on a screen using the technique called **touch sensing**.

30. With the use of **touch** sensing, computer displays can be tailored to the needs of the application rather than being limited by a fixed keyboard.

31. The TV-like display screen of a computer is called a **cathode ray tube (CRT)**.

32. With the use of an input device called a **light pen**, a user can write directly on a CRT screen.

33. The user can move a cursor around with a **joystick** device that is similar to the controllers used for video games.

34. CRTs are called **soft copy** devices because they do not produce paper documents.

35. A **hard** copy can be produced when a CRT is connected to a printer.

Enrichment Piece: Touch Tour

36. The lists of topics that are displayed on touch sensing screens are called **menus**.

37. Disney's EPCOT makes wide use of **touch** sensing terminals to help guide visitors through the park.

Other Input Devices

38. **Speech recognition** devices analyze spoken cammands and convert them to computer-stored data.

39. **TV** cameras can be used to facilitate the conversion of pictures into digital data.

40. A **membrane** keyboard has no movable keys.

Source Data Automation

41. The bypassing of manual data entry is called **source data automation (SDA)**.

42. The goal of **SDA** is to capture data automatically and in computer-readable form where it

originates.

43. The source data automation method used primarily in the banking industry is **magnetic ink recognition (MICR)**.

44. **MICR** is the SDA method used in the processing of checks.

45. **Optical character recognition (OCR)** is the SDA method in which the computer reads human-readable symbols directly.

46. **Optical mark recognition (OMR)** is the SDA technique popular in exam scoring.

47. Automated supermarket checkout systems use a form of **bar code** reading with the UPC.

Magnetic Ink Character Recognition

48. **MICR** reads characters written in magnetized ink.

49. **MICR** readers read the information that is imprinted on the bottom of a check and also microfilm the check as it passes through the reader.

Optical Character Recognition

50. **Optical character recognition** devices read handwritten and typewritten characters and symbols and convert them directly into computer codes.

51. A **font** is the complete assortment of letters, digits, and special characters of one style and size.

52. With a **hand**-print OCR reader, characters must be printed very carefully.

53. **Printed-font** OCR readers read typewritten and typeset characters.

54. **Omnifont** OCR readers can read most printed fonts and are generally the most expensive readers on the market.

55. The Kurzweil OCR reader is an example of an **omnifont** reader.

56. The most widely used OCR font used in the United States is **OCR-A**.

57. The OCR font people find more readable is **OCR-B**.

58. **Journal tape** readers are OCR devices that read printouts from cash registers.

59. **Small document** readers are OCR devices generally used to read credit card imprinted stubs.

60. The OCR readers that are capable of reading full pages of text, and sometimes in a variety of fonts are called **page** readers.

61. **Wand** readers are small hand-held units that bring OCR reading capability directly to the source of the data.

How OCR Works

62. The photocells in an OCR reader are read by a computer and digitized into a pattern of **bits**.

63. **Computer programs** inside the OCR reader analyze the bit patterns to determine which characters are being read.

OCR Applications

64. Information contained on **OCR** readable merchandise tags includes price, product code, and often a description of the item.

65. A **turnaround document** is printed by the computer, has additional information added, and is then resubmitted to the computer for processing.

Enrichment Piece: The Kurzweil OCR Reader

66. The **Kurzweil** OCR reader has a 40,000-word dictionary that allows it to check words with unrecognizable characters without the need for human intervention.

67. The Kurzweil OCR reader has been equipped with a **speech synthesizer** to develop a reading machine for the blind.

Optical Mark Recognition

68. With **optical mark recognition (OMR)**, pencil marks are recognized in specific positions on the page.

69. With **OMR** no stray pencil marks should be made on the test form.

Bar Code Reading

70. With **bar code** reading, coded information is read from a series of wide and narrow bands.

Enrichment Piece: Bar Code Applications

71. **Bar code** reading is a popular SDA technique being used to keep track of library books and speed the check-out of books.

Universal Product Code

72. The bar code convention adopted by food manufacturers is the **Universal Product Code (UPC)**.

73. The **UPC** code is now placed on most items sold in supermarkets.

74. The leftmost five digits of the **UPC** code identify the manufacturer.

75. The rightmost five digits of the **UPC** code identify the item.

76. The **UPC** coding scheme uses even parity; all representations of the digits must have an even number of ones.

77. To help determine if the bar code is read properly, UPC codes use even **parity**.

Benefits of UPC Scanning

78. With **UPC** scanning, supermarket check-out is faster and more accurate.

79. Scanning systems and computer systems can be used together to keep up-to-date records on inventory and prevent lost sales due to **stockouts**.

Controversial Issues in UPC Scanning

80. **Item pricing** is perhaps the most controversial issue surrounding UPC scanning.

Directions in UPC Scanning

81. Future UPC scanners will use laser **holography**, a technique for creating three-dimensional pictures, that will allow the scanners to read codes no matter how the item is positioned.

82. **Laser** holography is a technique that creates three-dimensional pictures.

Chapter 5
Output: Getting Results from Computers

Multiple Choice

1. The information we get from computers is called ______.
   (a) paging
   (b) input
   (c) ergonomics
   (d) output *
   (e) prompts

Printed Reports

2. The reports that list every item being processed are called ______ reports.
   (a) detail *
   (b) exception
   (c) ad hoc
   (d) demand
   (e) summary

3. ______ reports provide a concise overview of the information.
   (a) Exception
   (b) Summary *
   (c) Ad hoc
   (d) Detail
   (e) Periodic

4. The ______ report highlights information that needs human attention.
   (a) summary
   (b) internal
   (c) demand
   (d) detail
   (e) exception *

5. Reports that appear on a weekly schedule are called ______ reports.
   (a) ad hoc

(b) demand
(c) periodic *
(d) summary
(e) detail

6. ______ reports are regular reports that are produced as needed.
(a) Demand *
(b) Ad hoc
(c) Detail
(d) Summary
(d) Periodic

7. The ______ report is produced in response to a special request.
(a) periodic
(b) demand
(c) detail
(d) summary
(e) ad hoc *

8. ______ reports are sent to customers.
(a) Internal
(b) Ad hoc
(c) Demand
(d) External *
(e) Summary

9. A company will use carefully designed ______ forms to help convey a good impression.
(a) stock
(b) preprinted *
(c) laser
(d) summary
(e) microfiche

Types of Printers

10. Computer printers produce paper reports called ______.
(a) soft copy
(b) microfiche
(c) hard copy *
(d) holograms
(e) buffers

11. Printers operate as line printers, page printers, or ______ printers.
(a) dot-matrix
(b) impact

(c) nonimpact
(d) band
(e) character *

Impact Printers

12. ______ printers function like a typewriter and are capable of making multiple copies.
(a) Thermal
(b) Electrostatic
(c) Nonimapct
(d) Impact *
(e) Laser

13. Continuous form paper can be fed through a printer either by pin-feed or ______ feed.
(a) friction *
(b) scroll
(c) page
(d) form
(e) thimble

14. Printers that form characters by driving a series of pins against a ribbon are called ______ printers.
(a) daisy-wheel
(b) dot-matrix *
(c) formed character
(d) drum
(e) thimble

15. Daisy-wheel printers and thimble printers are examples of ______ printers.
(a) band
(b) chain
(c) formed character *
(d) line
(e) nonimpact

16. Because they produce print quality as good as an office typewriter, daisy-wheel and thimble printers are referred to as ______ printers.
(a) page
(b) nonimpact
(c) letter quality *
(d) line
(e) band

17. Band printers and ______ printers have their symbols arranged on a continuous loop.
(a) dot matrix

(b) thermal
(c) daisy-wheel
(d) thimble
(e) chain *

18. ______ printers may print lines that appear wavy.
(a) Band
(b) Thermal
(c) Daisy-wheel
(d) Drum *
(e) Chain

Nonimpact Printers

19. ______ printers use electronic or photographic techniques.
(a) Nonimpact *
(b) Dot matrix
(c) Impact
(d) Formed character
(e) Chain

20. Dot-matrix printers that have heated pins and use special heat-sensitive paper are ______ printers.
(a) electrostatic
(b) laser
(c) thermal *
(d) thimble
(e) formed character

21. Page printers that use light or laser beams to charge a metal drum are ______ printers.
(a) thermal
(b) electrostatic *
(c) formed character
(d) impact
(e) chain

22. ______ are the fastest nonimpact printers.
(a) Chain
(b) Thermal
(c) Band
(d) Daisy-wheel
(e) Laser *

23. The ability of a printer to print on both sides of a page is called ______.
(a) paging

(b) page buffering
(c) scrolling
(d) duplexing *
(e) reversed fields

Terminal-Oriented Systems

24. The visual display on a terminal is called ______.
(a) duplexing
(b) hard copy
(c) soft copy *
(d) page buffer
(e) ergonomics

25. The dialogue between the computer and the user is referred to as ______ computing.
(a) voice response
(b) conversational *
(c) talkative
(d) phoneme
(e) linear predictive

26. The message that appears on the screen to which the user responds is called a ______.
(a) cursor
(b) protected field
(c) self-test
(d) prompt *
(e) conversor

Cathode Ray Tube Terminals

27. On a CRT, a beam of light strikes a special coating called a ______.
(a) phosphor
(b) cursor home
(c) prompt
(d) scan assembly
(e) oscilloscope

28. CRTs work in the same manner as a ______.
(a) printer
(b) laser
(c) TV *
(d) laser hologram
(e) acoustic coupler

How a Laser Printer Works

29. The ______ on a laser printer forms the actual print image as a pattern of "on" and "off" bits.
    (a) input/output buffer
    (b) page buffer
    (c) scan assembly
    (d) laser printhead assembly
    (e) character generator *

30. The ______ on a laser printer is where an electronic image of the page is created as a pattern of "on" and "off" bits.
    (a) scan assembly *
    (b) input/output buffer
    (c) character generator
    (d) page buffer
    (e) charged drum

Features of CRT Terminals

31. The feature of a CRT that displays the most current lines of text while moving the oldest ones off the screen is called ______.
    (a) underscoring
    (b) paging
    (c) split screen
    (d) scrolling *
    (e) character or line blinking

32. The feature of a CRT terminal that allows the user to display a prior screen image is ______.
    (a) split screen
    (b) paging *
    (c) underscoring
    (d) reversed fields
    (e) partial screen transmit

33. The feature of a CRT terminal that allows one area to be scrolled while another area remains fixed is called ______.
    (a) reversed fields
    (b) protected fields
    (c) split screen *
    (d) underscoring
    (e) partial screen transmit

34. The ______ feature of a CRT terminal displays dark characters on a light background.
    (a) character or line brightness control
    (b) split screen
    (c) protected fields
    (d) underscoring
    (e) reversed fields *

35. The feature of a CRT terminal that draws the user's attention to a line of text is the ______.
    (a) partial screen transmit
    (b) scrolling
    (c) protected field
    (d) character or line blinking *
    (e) split screen

36. ______ are information areas that cannot be changed on a CRT screen.
    (a) Protected fields *
    (b) Reversed fields
    (c) Scrolled fields
    (d) Split screens
    (e) Partial screen transmit

37. The feature of a CRT terminal that works with the protected field feature is ______.
    (a) cursor home
    (b) underscoring
    (c) partial screen transmit *
    (d) reversed fields
    (e) scrolling

38. The feature that allows characters to be underlined on the CRT screen is ______.
    (a) reversed fields
    (b) underscoring *
    (c) character or line blinking
    (d) character or line brightness control
    (e) split screen

39. The feature of a CRT that sends the blip of light that signals your position on the screen back to its starting position is ______.
    (a) cursor base
    (b) underscoring
    (c) cursor control
    (d) character or line blinking
    (e) cursor home *

40. The internal program that a CRT can run to determine if it functioning properly is called ______.
    (a) underscoring

(b) cusor control
(c) self test *
(d) character generator
(e) scan assembly

Ergonomics

41. The study of the relationships between humans and machines is called human engineering or ______.
(a) peoplemation
(b) anthropomorphics
(c) duplexing
(d) ergonomics *
(e) humanomics

Portable Terminals

42. The built-in device on a portable terminal that allows the terminal to communicate over phone lines is a(n) ______.
(a) voice response system
(b) acoustic coupler *
(c) speech synthesizer
(d) speech recognition device
(e) oscilloscope

43. Many people can conduct business outside their offices by using ______.
(a) portable computer terminals *
(b) voice response systems
(c) phoneme coding
(d) computer output microfilm
(e) microfiche

The World of Computer Graphics

44. The creation and manipulation if pictures by a computer is called computer ______.
(a) holograms
(b) output microfilm
(c) graphics *
(d) ergonomics
(e) assisted image processing

45. ______ shoot colors onto paper through nozzles.
(a) Dot-matrix printers

(b) Drum plotters
(c) Multipen plotters
(d) Electrostatic plotters
(e) Ink-jet printers *

46. The process of using a computer to sharpen satellite photos is called ______.
(a) phoneme coding
(b) holography
(c) duplexing
(d) image enhancement *
(e) color graphics

Other Types of Computer Output

Computer Output Microfilm

47. ______ is a method of storing computer data that does not need to be read back into a computer.
(a) Dry-silver film processing
(b) Computer output microfilm *
(c) Duplexing
(d) Computer graphics
(e) Phoneme coding

48. ______ is used with COM systems to locate and display images quickly.
(a) Waveform digitization
(b) Phoneme coding
(c) Computer-assisted image processing *
(d) Conversational computing
(e) Duplexing

49. Many COM systems output information onto ______ capable of holding hundreds of pages of output.
(a) microfiche *
(b) phonemes
(c) computer graphics
(d) holograms
(e) cards

50. Which of the following is not a disadvantage of microfiche? ______
(a) file integrity
(b) can only be read with magnifying devices
(c) cannot be modified
(d) security
(e) compact *

Speech Synthesis

51. Computers can speak their output through a device called a speech ______.
    (a) compactor
    (b) synthesizer *
    (c) phoneme
    (d) oscilloscope
    (e) coupler

52. "The number you have just dialed has been changed to .....," is an example of a ______.
    (a) phoneme
    (b) microfiche
    (c) voice response *
    (d) waveform digitization
    (e) linear predictive coding

53. Two problems with various types of speech synthesizers today are the mechanical sounding voice and ______.
    (a) response bifurcation
    (b) oscilloscoping
    (c) digitization distortion
    (d) limited vocabulary *
    (e) prompt time

Speech Synthesis Techniques

54. ______ works by sampling the human voice.
    (a) Voice response
    (b) Phoneme coding
    (c) Linear predictive coding
    (d) Computer-assisted image processing
    (e) Waveform digitization *

55. With ______, common sounds in the English language are combined to enable a computer to talk.
    (a) voice response systems
    (b) phoneme coding *
    (c) hologram enunciation
    (d) linear predictive coding
    (e) waveform digitization

56. In ______, mathematical formulas are used to produce speech.
    (a) holograms

(b) voice response systems
(c) linear predictive coding *
(d) phoneme coding
(e) waveform digitization

The Shroud of Turin: An Ancient Mystery

57. Using a computerized image analyzer, scientists studying the Shroud of Turin discovered that the Shroud contained _____ information.
(a) three-dimensional *
(b) two-dimensional
(c) holographic
(d) phoneme
(e) waveform

Three-Dimensional Computer Output

58. Three-dimensional imagess produced by computers are called _____.
(a) graphics
(b) fiche
(c) phonemes
(d) holograms *
(e) phosphors

Applications for the Future

59. With the use of _____ pilots will be able to learn how to fly a plane without leaving the ground.
(a) voice response systems
(b) speech synthesis
(c) microfiche
(d) waveform digitization
(e) holographic simulators *

True/False

Printed Reports

1. Detail reports list every item being processed. (T p102)

2. Summary reports highlight individual items requiring human attention. (F p102)

3. Exception reports provide a concise overview of the information. (F p102)

4. Summary and exception reports are particularly useful to busy managers. (T p102)

5. Demand reports generally represent the bulk of an organization's reporting. (F p102)

6. Ad hoc reports are designed and produced in response to special requests. (T p102)

7. Ad hoc reports are essential in decision support systems. (T p102)

8. Internal reports of an organization are generally sent to that organization's customers. (F p102)

9. External reports are carefully designed to be readable and understandable; internal reports can appear cluttered. (T p102)

10. Stock forms are more "general-purpose" forms than are preprinted forms. (T p102)

11. One advantage of stock forms is that they can be attractively designed to help a company convey a good impression to its clients. (F p102)

Types of Printers

12. Page printers generally form a line at a time but print a page at a time. (T p102)

Impact Printers

13. Nonimpact printers can print multiple copies by using special forms with carbon paper. (F p104)

14. A KSR device can receive information from the control computer and it can transmit information entered at its keyboard to the central computer. (T p105)

15. Daisy wheel printers are character printers. (T p108)

16. Daisy wheel printers and thimble printers are examples of letter-quality printers. (T p108)

17. Dot-matrix printers are generally considered to be formed character printers. (F p108)

18. Correspondence quality printers produce letters that are sharper than those produced by letter quality printers. (F p109)

19. Band printers are generally nonimpact printers. (F p109)

20. Drum printers are line printers with characters arranged in circles around a cylindrical drum. (T p110)

21. One problem with drum printers is that the print lines may appear wavy. (T p110)

Nonimpact Printers

22. Impact printers use thermal, electrostatic, and laser methods to create their images. (F p110)

23. Laser printers generally form characters out of precisely placed dots. (T p110)

24. Laser printers produce multiple copies of reports by using special carbon papers. (F p110)

25. Laser printers can print grapics as well as text. (T p 111)

Terminal-Oriented Systems

26. Terminals generally operate on-line to the computer. (T p111)

27. With on-line, terminal-oriented systems, the computer types messages to the user; these are called prompts. (T p111)

Cathode Ray Tube Terminals

28. In a CRT terminal, a heater coil heats up a deflection structure causing it to emit a beam of electrons. (F p112)

29. In a CRT terminal, characters are formed by having electrons strike a phosphor coating that then glows. (T p112)

Features of CRT Terminals

30. The paging feature allows a CRT terminal user to redisplay prior screen information or to "skip ahead" a full line at a time. (F p114)

31. With the split screen CRT feature, some areas can be scrolled while others remain fixed. (T p114)

32. With the partial screen transmit feature of a CRT, only data recently entered and protected fields are transmitted to the CPU. (F p114)

33. The curse is a blip of light that may be moved to any position on a screen to call attention to that location. (F p114)

34. The cursor home function normally returns the cursor to the top right of a screen. (F p116)

How a Laser Printer Works

35. A laser printer works by charging an entire drum, and then shining a laser beam at selected points (to discharge these areas); toner then sticks to only the charged areas. (T p112)

Ergonomics

36. Item pricing is an ergonomics issue. (F p115)

37. An important ergonomic issue is whether or not CRTs are health hazards. (T p115)

Portable Terminals

38. Many portable terminals have built-in acoustic couplers, devices that enable the terminals to communicate with computers over standard telephone lines. (T p116)

The World of Computer Graphics

39. The three ink colors commonly used in ink-jet printers are yellow, magenta, and cyan. (T p118)

40. With multipen plotters, several pens may be drawing simultaneously. (F p120)

41. Plotters generally use static electricity to hold a document down while plotting. (T p120)

42. Plotters are particularly useful for obtaining hard copies of soft copy screen presentations. (T p120)

43. The Videoprint device makes high quality off-the-screen photographs. (F p121)

44. Computer graphics techniques can greatly enhance the features in a photograph. (T p123)

Computer Output Microfilm

45. Wet-silver film processing, neater than the old dry-chemical film processing, helps make the use of COM appealing. (F p124)

46. Microfiche are rolls of microfilm that can each hold hundreds of pages of computer output. (F p124)

47. One great advantage of computer output microfilm on fiche is that duplicates can be made quickly and economically. (T p124)

Speech Synthesis

48. The quality of speech available in most commercial products is generally considered to be about as good as human speech. (F p125)

Speech Synthesis Techniques

49. The sound spectrograph helps speech engineers to "visualize" speech. (T p126)

50. With waveform digitization, speech waves are sampled periodically, and the condensed information is then played back; the played back voice is as "rich" as the original. (F p126)

51. To form all possible English words requires between 400 and 700 phonemes. (F p126)

52. With linear predictive coding, speech is formed by solving mathematical equations to produce waveforms that are then played through speakers. (T p126)

Three-Dimensional Computer Output

53. Computer-controlled devices can produce three-dimensional outputs by the technique of laser holography. (T p116)

Fill-in-the-Blanks

1. Information from computers is known as **output**.

Printed Reports

2. **Detail** reports list every item being processed.

3. A concise overview of information is provided in **summary** reports.

4. **Exception** reports highlight individual items requiring human attention.

5. Every item that a store carries would be listed in a **detail** report.

6. A **summary** report might list total sales by department.

7. An **exception** report might list weak selling items.

8. **Detail** reports tend to be used by company personnel in their day-to-day responsibilities.

9. Busy managers find that **summary** and exception reports are especially helpful.

10. Reports that appear on a regular basis are called **periodic** reports.

11. A report that appears weekly, monthly, quarterly, or annually is a **periodic** report.

12. **Demand** reports are produced as needed.

13. **Ad hoc** reports are designed and produced in response to special requests.

14. **Ad hoc** reports are essential to decision support systems.

15. **Internal** reports are generated for use within an organization.

16. Reports sent to customers or other companies are called **external** reports.

17. **Stock** forms are generally blank or lined forms on which any report can be printed.

18. **Preprinted** forms generally contain headings and ruled lines.

Types of Printers

19. The output device for producing reports on paper is the **printer**.

20. Printers can be either impact or **nonimpact**.

21. **Character** printers print only one character at a time.

22. **Line** printers print an entire line at once.

23. **Page** printers form one line at a time, but actually print an entire page at a time.

Impact Printers

24. **Impact** printers function like typewriters.

25. **Impact** printers strike a mechanical hammer against a ribbon to leave an impression on paper.

26. It is costly to **decollate** (separate) multipart paper.

27. **Continuous** form paper is usually fanfolded and perforated.

28. Paper can be advanced through a printer by either a friction feed mechanism or a **pin-feed** mechanism.

29. **Dot-matrix** impact printers create an image by driving a series of pins against a ribbon.

30. The two common types of formed character printers are **daisy-wheel** and thimble printers.

31. Formed character printers are also called **letter-quality** printers.

32. With **band** printers, characters and symbols are arranged on a continuous loop.

33. Text printed on **drum** printers may appear to be wavy.

Nonimpact Printers

34. **Nonimpact** printers generally use electronic or photographic techniques to produce reports.

35. **Thermal** printers use heated pins and heat-sensitive paper.

36. Images printed with **thermal** printers can tend to fade.

37. **Electrostatic** printers generally use light beams or laser beams to induce electric charges on paper or a metal drum.

38. The fastest nonimpact printers available are **laser** printers.

39. The ability of some printers to print on both sides of a page is called **duplexing**.

Terminal-Oriented Systems

40. In **conversational** computing, the user carries on a dialogue with the computer.

41. The message the computer types to an interactive user is called a **prompt**.

Cathode Ray Tube Terminals

42. TV-like **cathode ray tube (CRT)** terminals are commonly used in timesharing applications.

43. The special coating on a **CRT** screen that glows when struck by a beam of electrons is called a phosphor.

Enrichment Piece: How a Laser Printer Works

44. The temporary storage area of a laser printer where information to be printed is held is called the **input/output** buffer.

45. The processor of the laser printer formats a complete page at a time and passes the page image to a storage area called the **page** buffer.

46. In a laser printer, the **character generator** forms the actual print image of each character as a pattern of "on" and "off" bits.

47. An electronic image of a page is created in the **scan assembly** of a laser printer.

48. The scan assembly of a laser printer sends the "on" and "off" bits to the **laser printhead assembly**.

Features of CRT Terminals

49. The feature of a CRT that moves the oldest line of text off the top of the display screen to make room to display the next line is called **scrolling**.

50. **Paging** allows a terminal user to redisplay prior screen images or to "skip ahead" a full page at a time.

51. The feature of a CRT that allows the screen to be divided into several separate areas is called **split screen**.

52. With **reversed fields**, information can be highlighted on the CRT screen as light letters on a dark background if the normal displays are dark letters on a light background.

53. **Character or line blinking** is a CRT feature which causes a character, a group of

characters, or an entire line to flash repeatedly.

54. The **protected field** feature of a CRT prevents certain information that appears on the screen from being changed.

55. The **partial screen transmit** is a feature of a CRT that works with the protected field feature.

56. The **character or line brightness control** is a feature of a CRT that lets certain characters or lines be displayed more brightly than others.

57. The blip of light that appears on the CRT screen which indicates the user's position is called a **cursor**.

58. **Cursor control** will move the cursor left, right, up, or down on the screen.

59. A **cursor home** function is often provided to return the cursor to its starting position.

60. The **self-test** feature of a CRT allows the user to cause an internal program to be run to determine if the CRT is functioning properly.

Enrichment Piece: Ergonomics

61. The relationship between humans and machines is called **ergonomics**, or human engineering.

62. Improving the person-machine relationship is one of the most important goals of **ergonomics**.

Portable Terminals

63. With **portable** computer terminals it is possible for people to conduct business anywhere outside their offices.

64. The device that attaches to the headset of a telephone and allows a computer to communicate over standard telephone lines is an **acoustic coupler**.

Enrichment Piece: The World of Computer Graphics

65. The creation and manipulation of pictures by computer is called **computer graphics**.

66. The use of **ink-jet** printers causes drops of colored ink to shoot out of a nozzle onto paper.

67. A **drum plotter** is a device that uses several different colored pens to trace out a picture as a sheet of paper is rolled back and forth.

Other Graphic Output Devices

68. With a **desktop** plotter it is easy to get a precise copy of a soft copy image from a CRT screen.

Computer-Aided Design (CAD)

69. **Computer-aided design (CAD)** is a popular means of using computers in designing buildings and electronic circuits.

Image Enhancement

70. **Image enhancement** is the term for using computer processing to sharpen pictures.

Other Types of Computer Output

Computer Output Microfilm

71. **Computer output microfilm (COM)** has become a popular way of storing hard copy computer output without producing paper copies.

72. Many COM users prefer **dry-silver** film processing rather than wet-chemical film processing.

73. With **computer-assisted image processing**, any image stored on microfilm can be located and displayed in seconds.

74. The 4-inch by 6-inch film cards produced by COM equipment are called **microfiche**.

75. Because it is easy to remove a fiche card from a file, fiche are said to lack **file integrity**.

Speech Synthesis

76. A computers can speak its outputs through a **speech synthesizer**.

77. Computers that answer telephones use what are called **voice response** systems.

78. The telephone company often uses computerized **voice response** systems to deliver

messages such as "The number you have called, ..., has been changed to ..."

Speech Synthesis Techniques

79. The speech synthesis technique that involves studying the human voice on a oscilloscope is **waveform digitization**.

80. The common sounds of the English language are called **phonemes**.

81. With the **phoneme coding** speech synthesis technique, the common sounds of the English language are combined to enable a computer to speak.

82. The speech synthesis method that uses a mathemetical approach is **linear predictive** coding.

Three-Dimensional Computer Output

83. Three-dimensional outputs that are produced by computer-controlled lasers are called **holograms**.

84. With a laser **hologram**, a person can walk around the picture and see an object that isn't even there.

Enrichment Piece: The Shroud of Turin: An Ancient Mystery

85. Computers are being used to study the **Shroud of Turin**, believed to be the burial cloth of Jesus of Nazareth.

86. By using a computerized image analyzer, scientists made a **three-dimensional** relief image of the Shroud of Turin appear on national television.

Applications for the Future

87. **Three-dimensional** holographic flight simulators may soon be used to teach pilots how to fly new planes while still on the ground.

Chapter 6
Secondary Storage

Multiple Choice

1. Disks and tapes are examples of ______.
   (a) facsimile storage
   (b) magnetic bubble memory
   (c) channels
   (d) density
   (e) auxiliary storage *

2. Another name for auxiliary storage is ______ storage.
   (a) flexible
   (b) secondary *
   (c) instant
   (d) bubble memory
   (e) density

Magnetic Tape Storage

3. ______ is made from a strong lightweight plastic and commonly wound on 10 1/2-inch reels.
   (a) Punched tape
   (b) Cassette tape
   (c) Sequential tape
   (d) Magnetic tape *
   (e) Seek tape

4. Two applications for magnetic tape are batch-processing and ______.
   (a) backing up *
   (b) formatting
   (c) densitizing
   (d) initializing
   (e) seeking

5. The ______ is capable of taking information off a tape and placing data onto a tape.
   (a) track head
   (b) channel head
   (c) read/write head *

(d) sectorhead
(e) block head

6. The number of bytes per inch that can be recorded onto tape is called ______.
   (a) density *
   (b) track size
   (c) block size
   (d) physical record size
   (e) logical record size

7. The ______ bit is an error detection device that checks for coding errors on tape.
   (a) density
   (b) channel
   (c) seek
   (d) parity *
   (e) character

8. The are commonly nine channels or ______ on the tape.
   (a) characters
   (b) tracks *
   (c) blocks
   (d) cylinders
   (e) sectors

9. Tracks on tape are also called ______.
   (a) sectors
   (b) cylinders
   (c) characters
   (d) blocks
   (e) channels *

10. EBCDIC and ASCII are examples of ______.
   (a) parity bits
   (b) transaction files
   (c) character sets *
   (d) direct organization
   (e) direct access

Transfer Rate

11. The number of characters per second that can go from main storage to a secondary storage device is referred to as the ______ rate.
   (a) data transfer *
   (b) instant access
   (c) latency

(d) random access
(e) seek

Organizing Records on Tape

12. Records are kept in order on tape for what is called ______.
(a) random access
(b) direct access
(c) instant access
(d) sequential access *
(e) recording access

13. The section of tape that is kept empty is called the ______.
(a) block
(b) interblock gap *
(c) blocksize
(d) blocking factor
(e) sector

14. To minimize the number of interblock gaps needed, records are recorded in groups or ______.
(a) channels
(b) tracks
(c) blocksizes
(d) seeks
(e) blocks *

15. If there are 20 records in a block and each has 80 characters, multiplying these figures yields the ______.
(a) logical record size
(b) tracks
(c) blocksize *
(d) blocking factor
(e) interblock gap

16. The blocksize is the same as the ______.
(a) physical record size *
(b) interblock gap
(c) logical record size
(d) block
(e) record

Disk Storage

17. The most popular secondary storage medium in instant access systems is ______.
    (a) magnetic tape
    (b) punched paper tape
    (c) bubble memory
    (d) disk *
    (e) facsimile

18. ______ consist of several flat platters stacked one on top of the other and connected by a spindle.
    (a) Facsimile storages
    (b) Disk packs *
    (c) Master files
    (d) Cylinder-track-and-records
    (e) Diskettes

19. Information is recorded on disks in circles called ______.
    (a) seeks
    (b) bubbles
    (c) cylinders
    (d) channels
    (e) tracks *

20. When the read/write head jumps between tracks on a disk we have what is called a ______.
    (a) channel
    (b) cylinder
    (c) seek *
    (d) latency
    (e) collision

Floppy Disks

21. Another name for floppy disk is ______.
    (a) Winnie
    (b) optical disk
    (c) cassette
    (d) flexible disk *
    (e) platter

22. Information is recorded on floppy disks in circular tracks that are divided into ______.
    (a) sectors *
    (b) blocks
    (c) platters
    (d) cylinders
    (e) channels

23. ______ disks have sectors physically marked by a series of holes near the center.
(a) Optical
(b) Hard-sectored *
(c) Sequential
(d) Soft-sectored
(e) Winchester

24. Soft-sectored disks have sector locations magnetically recorded in a process called ______.
(a) sequential organizing
(b) hash crashing
(c) data transfer
(d) initializing *
(e) direct organizing

25. Another term for initializing a disk is ______.
(a) organizing
(b) latency
(c) data transfer
(d) hash crashing
(e) formatting *

26. Before floppy disks were available for personal computers, ______ were used.
(a) facsimile storage devices
(b) bubble memories
(c) Winnies
(d) cassette tapes *
(e) optical disks

Winchester Disks

27. ______ are permanently sealed in plastic coatings.
(a) Optical disks
(b) Winchester disks *
(c) Facsimile storage devices
(d) Floppy disks
(e) Bubble memories

28. ______ were developed to improve removable disk technology.
(a) Bubble memories
(b) Floppy disks
(c) Winchester disks *
(d) Optical disks
(e) Cassette disks

How Data is Stored on Disk

29. On direct access devices, data is generally accessed with a method called ______.
    (a) cylinder-track-and record *
    (b) parity access
    (c) channel access
    (d) latency access
    (e) bubble access

30. A stack of tracks is also called a ______.
    (a) bubble
    (b) disk pack
    (c) seek
    (d) channel
    (e) cylinder *

31. The time it takes to access a record of information has three components--seek time, latency time, and ______ time.
    (a) recording
    (b) sector
    (c) data transfer *
    (d) initializing
    (e) density

32. The time it takes to position the boom to the appropriate cylinder is called ______.
    (a) seek time *
    (b) access time
    (c) data transfer time
    (d) latency time
    (e) record time

33. The time it takes for a desired record to spin around to the read/write head is called ______ time.
    (a) access
    (b) latency *
    (c) seek
    (d) data transfer
    (e) cylinder

34. The time it takes for a complete record to pass under the read/write head is called the ______ time.
    (a) latency
    (b) cylinder
    (c) seek
    (d) access
    (e) data transfer *

Mass Storage Devices

35. Many companies maintain large ______ that can be automated with the use of mass storage devices.
    (a) files
    (b) records
    (c) tape libraries *
    (d) random access files
    (e) direct access files

36. The IBM 3850 mass storage system stores tapes in a unit that resembles a ______.
    (a) library
    (b) honeycomb *
    (c) bookshelf
    (d) file cabinet
    (e) file folder

Bubble Memory

37. Bits stored as moving magnetic spots on synthetic garnet are called ______.
    (a) records
    (b) blocks
    (c) channels
    (d) bubbles *
    (e) tracks

38. ______ memory was developed at Bell Laboratories and will not lose data if electrical power is turned off.
    (a) Facsimile
    (b) Disk
    (c) Bubble *
    (d) Random access
    (e) Optical

Optical Disks

39. A type of storage that does not use magnetic recording principles is ______.
    (a) optical disk *
    (b) bubble memory
    (c) tape
    (d) floppy disk
    (e) Winnies

40. Bits are placed on optical disks with a ______.
    (a) electric current
    (b) laser *
    (c) read/write head
    (d) bubble generator
    (e) coil

41. Optical disks that can not be changed are used for ______.
    (a) bubbles
    (b) block storage
    (c) sequential storage
    (d) facsimile storage *
    (e) tape libraries

File Organization Methods

42. With the use of ______ organization, records can be accessed either sequentially or directly.
    (a) optical
    (b) batch
    (c) sector
    (d) direct
    (e) indexed sequential *

43. Records organized so that they may be accessed immediately are in ______ organization.
    (a) direct *
    (b) sequential
    (c) facsimile
    (d) batch
    (e) blocked

Sequential Files

44. Records are placed in order by ______ in sequential files.
    (a) master file
    (b) transaction file
    (c) name
    (d) key *
    (e) index

45. A ______ file might have a record for every item carried by a store indicating a part number, description, and the number on hand.
    (a) transaction

(b) master *
(c) direct
(d) detail
(e) sequential

46. A ______ file might contain all the adjustments to inventory including sales and arrivals.
(a) master
(b) sequential
(c) direct
(d) random
(e) detail *

47. A detail file is also called a ______ file.
(a) random
(b) master
(c) transaction *
(d) direct
(e) sequential

Direct Files

48. With ______ access a record on a disk can be accessed without a sequential search.
(a) random *
(b) sequential
(c) collision
(d) transaction
(e) key

49. Another name for random access is ______ access.
(a) sequential
(b) key
(c) collision
(d) transaction
(e) direct *

50. ______ is a technique for facilitating direct access.
(a) Collision
(b) Hashing *
(c) Hash crash
(d) Formatting
(e) Initializing

51. When two items hash into the same location, a ______ occurs.
(a) seek
(b) sector

(c) collision *
(d) block
(e) coincidence

52. Another name for a collision is a ______.
(a) coincidence
(b) block crash
(c) sector crash
(d) hash crash *
(e) index crash

Indexed Sequential Files

53. A(n) ______ contains an entry for every record key and its location.
(a) cylinder
(b) index *
(c) block
(d) mater file
(e) sector

Applications for the Future

54. Computer storage is becoming smaller, cheaper, faster, and more ______.
(a) bubbly
(b) user friendly
(c) energy efficient *
(d) magnetic
(e) sector resistant

True/False

Magnetic Tape Storage

1. Most of the tapes used on computer systems today are punched paper tapes. (F p134)

2. Half-inch computer tape is most commonly wound on 10 1/2-inch reels that hold 2400 feet of tape. (T p134)

3. Disk is commonly used for backing up (that is, making copies of) information recorded on tapes for security purposes. (F p134)

4. Tape normally moves 2000 inches per second on high-capacity tape drives. (F p135)

5. Tape densities of 1600 bpi are common. (T p135)

6. Today's fastest tape drives can process 125,000,000 characters per second. (F p135)

7. The two most common character sets are EBDCIC and ASCII. (F p136)

Transfer Rate

8. A 1600 bpi tape drive running at 200 inches per second has a data transfer rate of 320,000 cps. (T p136)

9. A 6250 bpi tape drive running at 200 inches per second has a transfer rate of 12,500,000 cps. (F p136)

10. Tape transfer rates of 5,000,000 cps are on the horizon. (T p136)

Organizing Records on Tape

11. Records are written and retrieved from tape in order one after the other in what is called sequential access. (T p137)

12. While the tape drive is accelerating, the read/write head passes over a section of tape called an intergap block. (F p137)

13. If 20 records are recorded together in one block, and each record requires 80 characters, the blocksize is 1600. (T p137)

14. If 20 records are recorded together in one block, and each record requires 80 characters, the logical record size is 1600. (F p137)

Disk Storage

15. Disk is the preferred secondary storage medium for today's instant access applications. (T p139)

16. When the disk's read/write head jumps between tracks this is called a seek. (T p140)

Floppy Disks

17. Winchester disks are sometimes called flexible disks. (F p140)

18. Information on a floppy disk is recorded in circular sectors divided into wedge-shaped

tracks. (F p142)

19. On soft-sectored disks, sector locations are magnetically recorded on the disk. (T p142)

20. Recording sector information is called formatting or initializing the disk. (T p142)

21. Before the invention of cassette tapes by Shugart Associates in 1972, personal computers used small floppy disks. (F p142)

Winchester Disks

22. Winchester disks are permanently sealed in plastic casings to prevent smoke and dust particles from damaging the surfaces. (T p143)

How Data is Stored on Disk

23. The "stack of tracks" traced out by the read/write heads on a disk forms a cylinder. (T p144)

24. Latency time is the time it takes for the complete record to pass under the read/write head. (F p145)

25. Seek time is the time it takes to position the boom to the appropriate cylinder. (T p145)

Mass Storage Devices

26. The IBM 3850 mass storage system was developed for automating massive tape libraries. (T p145)

27. The Automated Tape Library developed by the Braegen Corporation can locate and mount on a tape drive any of several thousand tapes in less than 20 seconds. (T p147)

Bubble Memory

28. Bubble memories are much faster than integrated circuits. (F p148)

29. It is believed that when the economics become more favorable, bubble memory may replace magnetic disk memory. (T p148)

Optical Disks

30. Optical disk technology does not use magnetic recording. (T p148)

31. Current optical disks that have pits burned into their surfaces by laser beams are not erasable. (T p148)

32. Optical disks that are read by laser beams are ideal for use in interactive educational applications. (T p149)

Sequential Files

33. In sequential files, records are always placed in order by part number. (F p150)

34. In an inventory control application, the transaction file would contain one record for each type of part. (F p150)

35. In an inventory control application, the master file would contain all the adjustments to the inventory including sales and arrivals. (F p150)

36. In an inventory control application, when a record appears on the master file but shows no current transactions, it simply means that this part was "inactive" in this processing period. (T p150)

37. In an inventory control application, if a record appears on the transaction file but not on the master file, there is a problem becaues information about "nonexistent" parts is being processed. (T p150)

Direct Files

38. Direct access allows the computer to access a certain record on a disk without a sequential search of every record before the desired one in the file can be found. (T p150)

39. One technique for facilitating sequential access is called hashing. (F p150)

40. In hashing schemes, the computer generally uses the same hashing calculation for both storing and retrieving records. (T p150)

41. Hashing schemes sometimes cause two different keys to hash into the same location on secondary storage. (T p150)

Indexed Sequential Files

42. Direct access is geared to applications in which massive amounts of data must be processed and a large portion of the file is processed on each run. (F p150)

43. Random access is geared to applications in which an individual piece of information must be located quickly, but only a small percentage of the file's records are processed in one session. (T p150)

44. Indexed sequential files are organized so that their records may be accessed either sequentially or randomly. (T p151)

45. The index of an indexed sequential file always contains one entry per record in the file. (F p151)

Applications for the Future

46. In general, the more information we have, the more informed the decisions we make. (F p151)

Fill-in-the-Blanks

1. External storage devices are known as **secondary** storage, or auxiliary storage.

2. Tape drivers and disk drivers are examples of **secondary storage** devices.

Magnetic Tape Storage

3. Holes punched into **punched paper** tape define the various character codes.

4. **Magnetic** tape is used on most large computer systems today.

5. **Magnetic** tape is made of a strong lightweight plastic that is commonly wound on 10 1/2-inch reels.

6. Tape is used in **batch** processing applications in which large amounts of data are processed on each run.

7. Tape can be used for security purposes by making copies; this process is called **backing up**.

8. The apparatus that can read or write data on tape is called the **read/write** head.

9. The number of characters that can be recorded per inch on tape is referred to as **density**.

10. **Characters** are represented on magnetic tape as a series of spots, each representing one bit.

11. A **parity** bit is an error detection device used to check for coding errors on tape.

12. There are commonly nine tracks, or **channels** on tape.

13. The bit patterns that correspond to each character belong to a **character** set.

14. The two most common character sets are **EBCDIC** and ASCII.

Transfer Rate

15. The number of characters per second that may be transferred between main storage and a secondary storage device is referred to as the **data transfer rate**.

16. The **data transfer rate** is determined by multiplying the density by the speed of the tape drive in inches per second.

17. The **data transfer rate** for a 1600-bpi tape drive running at 200 inches per second is 320,000 cps.

Organizing Records on Tape

18. Records are written and retrieved from tape in order one after the other in what is called **sequential** access.

19. The sections of the tape that are kept empty between blocks of recorded information are called **interblock gaps**.

20. Many logical records are recorded on tape in a group called a **block** to minimize the number of interblock gaps.

21. By multiplying the number of records in each block by the number of characters in each record we get what is called the **blocksize**.

22. If there are 20 records in a block, and each record requires 80 characters, the **blocking factor** is said to be 20.

23. Another name for the blocksize in the **physical record** size.

Disk Storage

24. **Disk** is the most popular secondary storage medium for instant access applications.

25. A disk **platter** resembles a phonograph record.

26. When several disks are stacked one on top of the other and connected by a spindle they form a **disk pack**.

27. Information is recorded on disks in circles called ***tracks***.

28. A **seek** is the process of the read/write head changing cylinders.

Floppy Disks

29. **Floppy disks**, sometimes called flexible disks or diskettes, are popular on personal computers.

30. Tracks on a floppy disk are divided into wedge-shaped **sectors**.

31. With **hard-sectored** disks, sector locations are physically marked by a series of holes near the center of the disk.

32. On **soft-sectored** disks, sector locations are magnetically recorded on the disk.

33. Recording of sector information on a disk is called **formatting** or initializing the disk.

34. Before the invention of the floppy disk, personal computers generally used **cassette tapes** for secondary storage.

Winchester Disks

35. Winchester disks are manufactured under "**clean** room" conditions.

36. **Winchester** disks are permanently sealed in plastic casings.

How Data Is Stored on Disk

37. Large computer systems generally access data on disk by a method known as **cylinder-track-and-record**.

38. The moving arm of a disk is called the **boom**.

39. A "stack of tracks" is called a **cylinder**.

40. **Seek** time is the time it takes to position the boom to the appropriate cylinder.

41. **Latency** time, or rotational delay, is the time it takes for the desired record to spin around to the read/write head.

42. The **data transfer** time is the time it takes for the complete record to pass under the read/write head.

Mass Storage Devices

43. Huge **tape** libraries are commonly replaced by mass storage devices.

44. The IBM 3850 honeycombed wall storage unit is an example of a **mass storage** device.

Bubble Memory

45. It is believed that bubble memory might eventually replace **disk** storage.

46. Bubble memories store bits as the presence or absence of magnetic **bubbles** on a thin film of synthetic garnet.

47. **Bubble** memories are stored on a thin flim of synthetic garnet.

48. Bubbles move along **tracks** etched in a garnet film.

49. New bubbles are created by a **bubble generator** coil.

Optical Disks

50. **Optical** disks do not use magnetic recording.

51. Information is recorded on optical disks using a pinpoint **laser beam** .

52. The surface of an optical disk has microscopic **pits** burned into it.

53. An optical disk is read by a **laser beam** that is too weak to damage the disk's surface.

54. Optical disks can be used only once because the pits are not **erasable** .

File Organization Methods

55. Records are organized one after the other in **sequential** organization.

56. In **direct** organization, records are organized so that any one may be referenced

immediately.

57. With **indexed sequential** organization, records are organized so that they may be read either sequentially or directly.

Sequential Files

58. In sequential files, records are placed in order by **key**.

59. A code that identifies a record is called a **key**.

60. A **master** file might contain one record for each item a manufacturer produces.

61. A **transaction** file, or detail file, might contain all the adjustments to the inventory including sales and arrivals.

Direct Files

62. Direct access is also called **random access**.

63. **Direct** access, or random access, allows access to a certain record on a disk without a sequential search of every record.

64. A technique for facilitating **direct** access is called hashing.

65. When two different account numbers hash into the same location on disk a **collision** or hash crash occurs.

Enrichment Piece: The Laser Card

66. Optical memory techniques allow color pictures, fingerprints and voice samples to be encoded onto **laser** cards.

Indexed Sequential Files

67. **Sequential** access is ideal for applications in which massive amounts of data must be processed and a large portion of the file is processed on each run.

68. **Direct** access is geared to applications in which an individual piece of information must be located quickly, and only a small percentage of the file's records are processed in one session.

69. **Indexed sequential** files are organized so that their records may be accessed either sequentially or randomly.

70. In indexed sequential files, the computer may maintain an **index** that lists the record key of every record.

71. An **index** may contain the record key and record location of every record.

Chapter 7
Data Communications

Multiple Choice

1. The transmission of computer information between computer systems and terminals is called ______.
   (a) telegraph communications
   (b) data communications *
   (c) private branch exchanging
   (d) analog dating
   (e) computerized branch exchanging

2. The trend today is toward ______ of cooperating computers that can share information.
   (a) concentrators
   (b) multiplexors
   (c) satellites
   (d) networks *
   (e) distributors

History

3. Samuel Morse developed the ______ that transmitted data as long and short tones.
   (a) telephone
   (b) modem
   (c) telegraph *
   (d) character set
   (e) simplex transmission

4. One of the earliest schemes for representing letters and numbers was ______.
   (a) Morse code *
   (b) ASCII
   (c) EBCDIC
   (d) analog data
   (e) digital data

5. The first patent for a device that could carry human voice was issued to ______.

(a) Paul Baran
(b) Samuel Morse
(c) Thomas Watson
(d) George Stibitz
(e) Alexander G. Bell *

6. ______ is the process by which a large central computer is shared by many users at once.
(a) Token passing
(b) Common carriers
(c) Timesharing *
(d) Polling
(e) Packet switching

7. ______ is the process of interconnecting many computers.
(a) Polling
(b) Networking *
(c) Timesharing
(d) Concentrating
(e) Common carriers

8. The United States agency whose rulings have greatly affected the data communications industry is the ______.
(a) PBX
(b) ISO
(c) ANSI
(d) FCC *
(e) CBX

How Data Communications Systems Work

9. Computers transmit bits of ______ data.
(a) analog
(b) fiber optic
(c) digital *
(d) node
(e) core

10. Telephones transmit ______ data.
(a) core
(b) node
(c) digital
(d) fiber optic
(e) analog *

11. The device that can transform digital data into analog data is called a ______.

(a) concentrator
(b) modem *
(c) stat mux
(d) common carrier
(e) multiplexor

12. A modem is generally connected to a ______ that handles many of the chores of managing data communications.
(a) private branch exchange
(b) central network control
(c) data communications controller *
(d) computerized branch exchange
(e) node

13. When terminals are connected to a computer system with standard telephones, modems with ______ are generally used.
(a) acoustic couplers *
(b) concentrators
(c) relay stations
(d) polling
(e) multiplexors

Asynchronous aand Synchronous Transmission

14. ______ transmission encloses individual characters between a start and stop bit.
(a) Synchronous
(b) Laser
(c) Cable
(d) Asynchronous *
(e) Microwave

15. ______ transmission is used with slow devices such as terminals.
(a) Microwave
(b) Asynchronous *
(c) Laser
(d) Cable
(e) Synchronous

16. With asynchronous transmission, a ______ bit and ______ bit are used.
(a) core, cladding
(b) node, polling
(c) start, stop *
(d) address, token
(e) packet, collision

17. ______ transmission handles characters as one continuous stream of bits.
(a) Laser
(b) Cable
(c) Asynchronous
(d) Microwave
(e) Synchronous *

18. The type of transmission that uses a clocking mechanism to recognize where bits start and stop is ______ transmission.
(a) synchronous *
(b) microwave
(c) laser
(d) asynchronous
(e) cable

Data Transmission Codes

19. The character set that was developed primarily for the communication of data between computers is ______.
(a) ANSI
(b) ISO
(c) EBCDIC
(d) ASCII *
(e) FDM

20. The two groups concerned with fostering standardization in the data-processing and data communications industries are ______ and ______.
(a) TDM, FDM
(b) ASCII, ISO
(c) ISO, ANSI *
(d) ANSI, ASCII
(e) FDM, ISO

Simplex, Half-Duplex, and Full-Duplex Transmission

21. The type of transmission that goes in one direction only is ______ transmission.
(a) full-duplex
(b) asynchronous
(c) half-duplex
(d) synchronous
(e) simplex *

22. ______ transmission allows transmissions in both directions but only one way at a time.
(a) Simplex

(b) Half-duplex *
(c) Asynchronous
(d) Full-duplex
(e) Synchronous

23. With ______ transmission, data can be transmitted in both directions simultaneously.
(a) synchronous
(b) simplex
(c) full-duplex *
(d) asynchronous
(e) half-duplex

24. The type of transmission generally used in transaction processing is ______ transmission.
(a) half-duplex *
(b) asynchronous
(c) full-duplex
(d) simplex
(e) syncchronous

25. The transmission that is in TV transmissions is ______ transmission.
(a) full-duplex
(b) asynchronous
(c) half-duplex
(d) simplex *
(e) synchronous

26. The type of transmission that is commonly used in CB radio transmission is ______ transmission.
(a) synchronous
(b) half-duplex *
(c) simplex
(d) asynchronous
(e) full-duplex

Line Speed

27. Transmission lines are classified according to the speed they carry data; this speed is called ______.
(a) datawidth
(b) pathwidth
(c) bandwidth *
(d) linewidth
(e) inflowidth

28. ______ lines were popular in the early telegraph system.

(a) Baseband
(b) Voiceband
(c) Broadband
(d) Leased
(e) Narrowband *

29. Fiber optic and microwave transmissions involve ______ lines.
(a) broadband *
(b) narrowband
(c) leased
(d) voiceband
(e) baseband

30. Line speeds are written in bits per second or ______.
(a) packets
(b) nodes
(c) polling
(d) baud *
(e) cores

Transmission Media

31. Which of the following consists of many pairs of twisted-pair wires? ______
(a) Cable *
(b) Coaxial cable
(c) Microwave transmission
(d) Fiber optic cable
(e) Satellite transmission

Twisted Pairs

32. The most popular medium for local telephone and data communications is currently ______.
(a) fiber optic glass cable
(b) twisted-pair wires *
(c) microwave transmission
(d) coaxial cable
(e) core cable

Cables

33. The medium that carries transmissions from central telephone exchanges to local telephones is generally ______.
(a) coaxial cable

(b) microwave transmission
(c) twisted-pair wires
(d) fiber optic glass cable
(e) cable *

Coaxial Cables

34. The medium that cable TV companies use to bring subscription television to subscribers is generally ______.
(a) coaxial cable *
(b) fiber optic glass cable
(c) microwave transmission
(d) satellite transmission
(e) twisted-pair wires

35. The electrical noise that is present in many transmission media is called ______.
(a) stat mux
(b) packets
(c) static *
(d) core bus
(e) token ring

Microwave Transmission

36. ______ transmission requires relay stations about 30 miles apart.
(a) Fiber optic
(b) Satellite
(c) Simplex
(d) Microwave *
(e) Full-duplex

Satellite Transmission

37. Communications satellites above the earth are placed into ______ orbits.
(a) typical
(b) geostationary *
(c) satellite
(d) constant
(e) common carrier

38. Satellite transmission also uses ______.
(a) fiber optics
(b) coaxial cables

(c) twisted-pairs
(d) lasers
(e) microwaves *

Fiber Optics

39. Fiber optic transmission sends ______ signals instead of electrical signals.
(a) light *
(b) digital
(c) analog
(d) baud
(e) sound

40. The source of light used in fiber optics is ______.
(a) claddings
(b) lasers *
(c) stat muxs
(d) microwaves
(e) packets

41. The core and the ______ in fiber optic cable are made of slightly different types of glass.
(a) outer coating
(b) packet
(c) cladding *
(d) node
(e) stat mux

42. Because ______ lines are not affected by electrical noise they are useful in factories where electrical generators operate.
(a) twisted-pair
(b) coaxial
(c) copper cable
(d) fiber optic *
(e) microwave

43. Because tampering with ______ is extremely difficult, it is popular with the military.
(a) satellite transmission
(b) fiber optic cable *
(c) microwave transmission
(d) coaxial cable
(e) full-duplex cable

The Purest Glass

44. The _____ used optical transmission of data.
   (a) telephone
   (b) photophone *
   (c) telegraph
   (d) photograph
   (e)photospeaker

Line Configuration

Point-to-Point and Multidrop Lines

45. On a _____ line, a terminal is directly connected to a computer with no other terminal on the line.
   (a) multidrop
   (b) multipoint
   (c) switched
   (d) multiplexed
   (e) point-to-point *

46. When several terminals share a line to the computer, we have a _____ line.
   (a) leased
   (b) point-to-point
   (c) multidrop *
   (d) switched
   (e) dedicated

47. When using a _____ line, only one terminal at a time can communicate with the computer, but the computer can broadcast a message to many terminals at once.
   (a) multidrop *
   (b) leased
   (c) point-to-point
   (d) switched
   (e) multiplexed

Leased and Switched Lines

48. Lines that are reserved for undivided attention are called _____ lines.
   (a) polled
   (b) leased *
   (c) multidrop
   (d) switched
   (e) multipoint

49. _____ lines establish a different line of communication each time a number is dialed.

(a) Multipoint
(b) Leased
(c) Point-to-point
(d) Switched *
(e) Multidrop

50. When a modem answers a call from a terminal, it sends a signal to the terminal that a connection has been established; this is called ______.
(a) collision concept
(b) polling
(c) packets
(d) handshaking *
(e) token passing

Polling with Multidrop Lines

51. When the communications controller tests to see if a terminal wishes to transmit to the computer, it is ______.
(a) addressing
(b) polling *
(c) token passing
(d) handshaking
(e) packet switching

52. Another name for a terminal or computer in a multidrop configuration is a ______.
(a) node *
(b) packet
(c) token
(d) address
(e) bus

Data Communication Networks

53. The ______ is one of the three most common network arrangements.
(a) local area network
(b) satellite network
(c) bus network *
(d) timesharing
(e) central network

Star Networks

54. In a ______ network a central computer communicates with various terminals and

computers over point-to-point lines.
(a) bus
(b) local
(c) ring
(d) star *
(e) distributed

55. ______ networks are commonly used in timesharing systems.
(a) Bus
(b) Distributed
(c) Ring
(d) Local
(e) Star *

Bus Networks

56. In a ______ network, if a node fails the entire system can continue to operate.
(a) token
(b) bus *
(c) local
(d) ring
(e) distributed

57. It is easy to add new nodes to a ______ network by tapping into the line.
(a) bus *
(b) distributed
(c) local
(d) star
(e) ring

Ring Networks

58. In ______ networks, token passing is used to determine which node transmits next.
(a) bus
(b) local
(c) ring *
(d) star
(e) distributed

59. A pattern of bits passed from node to node to determine which node transmits next is called ______.
(a) packet
(b) address
(c) contention

(d) token *
(e) poll

Local Networks

60. ______ are generally company owned and give management greater flexibility.
(a) Relay stations
(b) Satellites
(c) Star networks
(d) Local networks *
(e) Ring networks

61. Which of the following is not a local area network? ______.
(a) Wangnet
(b) Ethernet
(c) Optical highway
(d) Z-net
(e) Microwave *

Case Study 7-1: The Ethernet Local Network

62. The various pieces of office equipment on the Ethernet system are generally connected to ________.
(a) twisted-pair wires
(b) coaxial copper cable *
(c) fiber optic cable
(d) microwave transmission
(e) glass wires

63. Information on the Ethernet system travels in ______ over the cable.
(a) tokens
(b) nodes
(c) packets *
(d) polls
(e) lasers

64. Ethernet uses the ______ concept to determine which piece of equipment communicates next.
(a) collision *
(b) hash crash
(c) asychronous transmission
(d) synchronous transmission
(e) packet switching

65. When an individual device fails and the network continues to operate we generally have

_____ network control.
(a) central
(b) value-added
(c) packet
(d) distributed *
(e) contention

66. A _____ network can only transmit on a single channel.
(a) broadband
(b) satellite
(c) microwave
(d) local
(e) baseband *

67. _____ can handle both voice and data communications and operate over standard telephone lines.
(a) Timesharing
(b) PBX *
(c) Lasers
(d) Satellites
(e) Electronic mail

Multiplexors

68. A _____ may break a high-speed line into several lower speed lines.
(a) value-added service
(b) common carrier
(c) multiplexor *
(d) modem
(e) data communications controller

69. With _____ multiplexing, the line is dedicated to each terminal for brief periods.
(a) packet
(b) digital
(c) frequency-division
(d) time-division *
(e) analog

70. _____ works by assigning different frequency ranges to different terminals.
(a) TDM
(b) CBX
(c) FDM *
(d) PBX
(e) FCC

Statistical Multiplexors

71. A ______ gives unused line capacity to other terminals which are ready to transmit.
    (a) stat mux *
    (b) relay station
    (c) packet
    (d) node
    (e) token

Concentrators

72. ______ assemble many small transmissions into a single high-speed transmission to be sent over a single high-speed line.
    (a) Modems
    (b) Concentrators *
    (c) Acoustic couplers
    (d) Lasers
    (e) PBXs

Common Carriers

73. Companies authorized by the government to provide communications services to the public are called ______.
    (a) PBXs
    (b) value-added services
    (c) CBXs
    (d) common carriers *
    (e) concentrators

74. ______ was specifically responsible for greatly increasing competition in the data communications industry.
    (a) Carterphone *
    (b) Electronic mail
    (c) Packet switching
    (d) Mailgram
    (e) Teleconferencing

75. ______ is the value-added service that facilitates "face-to-face" meetings.
    (a) Packet switching
    (b) Teleconferencing *
    (c) Electronic mail
    (d) PBXs
    (e) CBXs

Expectations for the Future

76. Silicon chips will be replaced by chips made of ______.
    (a) mux
    (b) gallium arsenide *
    (c) bubble memory
    (d) copper
    (e) glass

True/False

History

1. Alexander Graham Bell observed, "If the presence of electricity can be made visible in any part of a circuit, I see no reason why intelligence may not be transmitted instantaneously by electricity. (F p156)

2. The first patent for a telephone that could carry human voices was issued to Samuel Morse in 1876. (F p156)

3. The International Morse Code symbol for SOS or "save our ship" is "... _ _ _ ..." (T p156)

4. In computer networking, many separate computers are interconnected. (T p157)

How Data Communications Systems Work

5. Analog data are transmitted as continuous waveforms. (T p157)

6. A modem transforms digital data to analog and vice versa. (T p157)

Asynchronous and Synchronous Transmission

7. Asynchronous transmission is used when high-speed transmission is required and when large groups of characters are normally ready for transmission at once. (F p158)

8. Asynchronous transmission handles individual characters by enclosing their bit patterns between start and stop bits. (T p158)

9. With asynchronous transmission, after a character is transmitted, the start bit for the next character is transmitted immediately. (F p158)

10. The hardware for asynchronous transmission is more costly than the hardware for synchronous transmission--this is because it must have precisely timed clocking mechanisms to recognize where each bit and each character begin and end. (F p158)

11. Analog transmission is preferred over digital transmission for long distance transmission. (T p159)

Data Transmission Codes

12. The two most popular character sets in use today are EBCDIC and ANSI. (F p160)

13. ASCII was developed by IBM for use in its System/360 line of computers. (F p160)

14. ASCII was developed primarly for communication of data between computer systems. (T p160)

15. EBCDIC represents letters, digits, and special symbols by variable-length patterns of bits. (F p160)

16. The International Standards Organization (ISO) and the American National Standards Institute (ANSI) are two groups concerned with fostering standardization in the data processing and data communications industries. (T p160)

Simplex, Half-Duplex, and Full-Duplex Transmission

17. With simplex transmission, data travels in only one direction; transmission in the opposite direction is not allowed. (T p160)

18. Half-duplex transmission is common in TV and one-way radio transmissions. (F p160)

19. Half-duplex transmission permits transmission in both directions but in only one direction at a time. (T p160)

20. Simplex lines are more costly than half-duplex lines. (F p160)

21. CB radio enthusiasts use simplex transmission. (F p160)

22. Full-duplex transmission sends data in both directions simultaneously. (T p160)

Line Speed

23. Narrowband lines operate up to 300 bits per second. (T p161)

24. Broadband lines operate between 300 and 9600 bps. (F p161)

25. Voiceband lines were popular in the early telegraph system. (F p161)

26. Human voice transmission requires about 2400 bps to be understandable. (T p161)

27. Narrowband lines are used in the telephone network. (F p161)

28. High-speed fiber optic communication lines would be considered broadband lines. (T p161)

29. Two of the most commonly used broadband transmission rates are 19,200 bps and 50K bps. (T p161)

Twisted Pairs

30. One of the original types of wire used in telephone communications is the twisted-pair wire. (T p161)

Cables

31. Cables consist of bundles of insulated twisted pairs. (T p161)

Coaxial Cable

32. Coaxial cable consists of a bundle of insulated twisted pairs. (F p161)

Microwave Transmission

33. Normally, microwave relay towers are placed about 300 miles apart. (F p162)

34. Microwave transmissions typically have high bandwidths. (T p162)

Satellite Transmission

35. Satellite transmission uses microwaves. (T p162)

36. Satellites are ordinarily maintained in geostationary orbits 222,300 miles above the surface of the earth. (F p162)

37. One disadvantage of satellite transmission in the superhigh frequency range is that

electronic noise in downtown city areas interferes with the signals. (F p164)

Fiber Optics

38. Copper cable has established itself as the communications cable technology of the future. (F p164)

39. Lasers produce "aligned" light beams. (T p164)

40. The primary function of the plastic outer coating on fiber optic cable is to force light beams to remain inside the cable. (F p164)

41. Fiber optic cable can actually bend a beam of light to follow the cable. (T p164)

42. Theoretically, a laser beam has the capability to carry as many as one billion telephone conversations. (T p164)

43. Today's half-inch copper cable can carry about 10 times as many conversations as today's fiber optic cable. (F p165)

44. One problem with fiber optic cable is that its transmissions are disturbed by electrical noise generated by nearby communication lines or electrical devices. (F p166)

45. Because fiber optic cables do not involve the transmission of electricity, grounding to prevent electrical shocks is not necessary. (T p166)

46. Fiber optic cables are particularly popular for high-security applications. (T p166)

Enrichment Piece: The Purest Glass

47. Samuel Morse invented the "Photophone" in 1880. (F p166)

Point-to-Point and Multidrop Lines

48. A terminal that is directly connected to a computer system with no other terminals on the line is said to use a point-to-point line. (T p166)

49. When several terminals share a single line connected to the computer, the line is called a multipoint line. (T p167)

50. In a multidrop configuration, the computer may send a message to only one terminal at a time. (F p167)

Leased and Switched Lines

51. Leased lines are dedicated lines. (T p167)

52. The public telephone network uses dedicated lines. (F p168)

53. On leased lines, communication is normally initiated by the sender dialing the receiver's number. (F p168)

Polling with Multidrop Lines

54. Dedicated leased lines are generally used in multidrop configurations. (T p168)

55. Each terminal or computer on a multidrop line is called a node. (T p168)

56. In polling, each time the computer checks with a particular terminal, the terminal transmits (or receives) its next block of information. (F p168)

57. In polling, each terminal is polled as often as all the others--this is done to ensure fairness. (F p168)

58. In polling, a polled terminal that indicates it wants to send data is given the uninterrupted attention of the central computer. (F p168)

59. Timer devices are commonly used with multidrop lines. (T p169)

Data Communications Networks

60. The most common network arrangements are star networks, bus networks, and ring networks. (T p169)

Star Networks

61. In a star network, a central computer communicates with various terminals and other computers over multipoint lines. (F p169)

62. In a star network, the various nodes (computers and terminals) are connected to the central computer but not to one another. (T p170)

63. In a star network, the central computer acts as a network controller. (T p170)

64. In a star network, if the central computer fails, the entire network fails. (T p170)

65. In star networks, polling is commonly used to determine which node will transmit next. (T p170)

66. Timesharing systems are generally ring networks with the central node being the timesharing "host" computer. (F p170)

Bus Networks

67. A bus network is essentially a single multidrop line shared by many nodes. (T p170)

68. In a bus network, the message to be transmitted is placed on the bus and is broadcast to all the nodes. (T p170)

69. In a bus network, the message being transmitted contains the address of the receiving node. (T p170)

70. In a bus network, if one node fails the entire network fails. (F p170)

71. In a bus network, new nodes may be added easily by "tapping" into the bus. (T p170)

Ring Networks

72. In a ring network, messages from one computer to another on the ring must be addressed to the destination computer. (T p170)

73. In ring networks, polling is commonly used to determine which node transmits next. (F p170)

74. In ring networks, each time a node receives the token, if the node is ready to transmit it transmits some data to the appropriate node, and then passes the token to the node to which it just transmitted the data. (F p170)

75. Ring networks are formed with point-to-point lines. (T p170)

76. In a ring network, if any node fails the entire network fails. (T p170)

Local Networks

77. The vast majority of business information is distributed within a very short distance from the office. (T p171)

78. Local networks are designed to handle communications within only a single building.

(F p171)

79. Local networks generally operate at slower speeds than the telephone network. (F p171)

80. One advantage of local networks is that they are generally independent of the public telephone system and its constraints. (T p171)

81. Xerox Corporation introduced the Z-net local network. (F p171)

Case Study 7-1: The Ethernet Local Network

82. Ethernet sends information as packets. (T p171)

83. Ethernet uses the collision concept to determine which node transmits next. (T p172)

84. In Ethernet, when one node has begun transmitting, all other nodes must wait. (T p173)

85. In Ethernet, if transmissions collide, all devices stop transmitting. (T p173)

86. The collision concept allows Ethernet to operate with central network control--this makes it easy to add new devices to the network after initial installation. (F 173)

87. Ethernet was initially designed to be a broadband network, but baseband versions are under development. (F p173)

88. A baseband network can only transmit on a single channel. (T p173)

89. Ethernet efficiently uses most of the capacity of its coaxial cable. (F p173)

90. Broadband systems generally have a much larger capacity than baseband systems, and they can transmit several channels at once. (T p173)

91. Wang's Wangnet is a broadband network. (T p173)

92. One problem with PBXs is that they can handle voice communications but not data communications. (F p173)

93. CBXs generally use distributed network control. (F p173)

94. CBX networks are generally cheaper to implement than coaxial cable networks. (T p173)

Multiplexors

95. A multiplexor divides the capacity of a high-speed line to make it appear to be several

low-speed lines. (T p174)

96. A multiplexor could allow four 300-bps terminals to share a 1200-bps line. (T p175)

97. FDM dedicates the line to each terminal for brief bursts of time. (F p175)

98. TDM permits a line to be shared by more terminals than FDM. (T p175)

99. FDM works by assigning different frequency ranges to different terminals. (T p175)

Statistical Multiplexors

100. A statistical multiplexor gives unused line capacity to other terminals which are ready to transmit. (T p176)

101. As long as any terminal is ready to transmit or receive, multiplexors will generally skip over inactive terminals and give the line capacity to active terminals. (F p176)

Concentrators

102. Unlike multiplexors, concentrators can "absorb" huge amounts of information and hold the data for long periods before transmitting it. (T p177)

103. Airline transportation is similar in concept to concentrator operation with data communications. (T p177)

Common Carriers

104. In the Carterphone decision, the FCC stated that messages could be put through networks only with devices manufactured, installed, or controlled by the network operators. (F p177)

105. The Carterphone decision brought many more competitors into the data communications industry. (T p177)

106. Teleconferencing and electronic mail are examples of value-added services. (T p178)

107. Packet switching was developed in the early 1960s in response to the U.S. government's fear of a possible nuclear attack by the Soviet Union. (T p178)

108. One goal of packet switching is to ensure virtually error-free transmission. (T p178)

109. A.T.&T. is the oldest communications company in the United States. (F p178)

110. Value-added carriers enable communication between many normally incompatiable devices. (T p178)

111. Gallium arsenide chips are expected to operate one billion times faster than today's silicon chips. (F p178)

112. It is believed that by the year 2000, computer power and communications capabilities will be about 1000 times greater than they are today--for about the same cost. (T p178)

Fill-in-the-Blanks

1. The name given to the process of transmitting computer data between computer systems and terminals is **data communications** .

2. The trend today is toward **networks** of cooperating computers that can share information.

History

3. **Morse** code was one of the earliest schemes for representing letters and numbers.

4. **Alexander Graham Bell** was issued the first patent for a telephone that could carry human voices.

5. With **timesharing** , a large central computer can be shared by many interactive users at once.

6. With **networking** , many computer systems are interconnected.

7. The **Federal Communications Commission (FCC)** is the government agency that issues the rulings that govern the communications industry.

How Data Communications Systems Work

8. Computers transmit bits of **digital** data.

9. Telephone lines transmit **analog** data.

10. The device that converts digital data into analog data is called a **modem** .

11. A **data communications** controller, is a device that receives signals from the

transmission line, converts them into computerized form, and passes them on to the central computer system.

12. The **data communications** controller handles many of the chores associated with managing data communications.

13. Modems with **acoustic couplers** can speak into a telephone headset.

14. **Modems** are normally capable of converting signals from digital to analog as well as from analog to digital forms.

Asynchronous and Synchronous Transmission

15. **Asynchronous** transmission is used with relatively slow devices such as terminals.

16. **Asynchronous** transmission encloses individual bit patterns between a start bit and a stop bit.

17. **Synchronous** transmission is used when high-speed transmission is required and when a large group of characters are normally ready for transmission at once.

18. With **synchronous** transmission there are no start bits or stop bits.

Enrichment Piece: Lincoln's Logic

19. The **telegraph** was used to send communications between battlefields during the American Civil War.

Data Transmission Codes

20. Of the two most popular character sets, EBCDIC and ASCII, **ASCII** was developed primarily for communications between computer systems.

21. The character set that was developed by IBM for its System/360 line of computers was **EBCDIC**.

Simplex, Half-Duplex, and Full-Duplex Transmission

22. Data travels in only one driection in **simplex** transmissions.

23. TV and one-way radio transmissions are examples of **simplex** transmission.

24. **Half-duplex** transmission allows transmission in both directions but only in one direction at a time.

25. **Transaction** -processing systems allow a user type a request to the computer, and the computer then processes the request and transmits the results to the user.

26. CB radio uses **half-duplex** transmission.

27. With **full-duplex** transmission, data can be sent in both directions simultaneously.

Line Speed

28. The transmission speed at which a transmission line can carry data is called **bandwidth** .

29. **Narrowband** , or low-speed, transmission lines operate up to 300 bits per second.

30. Medium-speed transmission lines are called **voiceband** lines.

31. The fastest transmission lines are **broadband** lines.

Transmission Media

32. A cable contains many **twisted-pair** wires.

Twisted Pairs

33. **Twisted-pair** wires were one of the original transmission media used in telephone communications.

Cables

34. Bundles of twisted pair wires are wrapped into a single large **cable** .

35. **Cables** are used to carry transmissions from central telephone exchanges to local telephones.

Coaxial Cables

36. **Coaxial** cables are used to bring multichannel TV programming into homes.

37. **Coaxial** cable is popular in local networking environments.

Microwave Transmission

38. With **microwave** transmission, relay stations are used to instead of cables and wires.

39. With **microwave** transmission, relay stations are used that are placed about 30 miles apart.

Satellite Transmission

40. **Satellite** transmission also uses microwave transmission.

41. Satellites are placed in a **geostationary** orbit above the surface of the earth.

Fiber Optics

42. The cable technology of the future is **fiber optics** .

43. With **fiber optics** , transmission is by light rather than electricity.

44. One of the most common sources of light used in fiber optics is **lasers** .

45. **Fiber optic** cable consists of a protective outer coating, cladding, and a core.

46. **Fiber optics** is immune to electrical disturbances.

Enrichment Piece: Laser Talk

47. A **laser beam** has the capacity to carry as many as 1 billion telephone conversations.

Enrichment Piece: The Purest Glass

48. Early optical transmission of data was done by a device invented by Alexander Graham Bell called the **Photophone** .

Line Configuration

Point-to-Point and Multidrop Lines

49. In a **point-to-point** line configuration, a computer terminal is directly connected to the computer system with no other terminal on the line.

50. With a **multidrop** or multipoint line configuration, several terminals share a single line to the computer.

51. In a **multidrop** line configuration, only one terminal at a time can transmit data to the computer.

Leased and Switched Lines

52. **Leased** lines are generally dedicated lines reserved for use by a particular party who is charged for the undivided attention of the line.

53. **Switched** lines are available through the regular telephone switching system.

54. With **leased** lines, a terminal is always connected to the computer.

55. With **switched** lines, a terminal user must dial the computer in the same manner as if dialing someone on a conventional telephone.

56. **Handshaking** occurs when a modem sends a signal down the line to indicate to the terminal that the connection has been established.

Polling with Multidrop Lines

57. Each terminal in a multidrop line is called a **node**.

58. In **polling**, the communications controller tests each terminal to determine if that terminal wishes to transmit data.

Data Communications Networks

59. It has become common for computers to communicate with one another over **data communication** networks--systems with more than one computer and terminal.

Star Networks

60. In a **star** network configuration, computers and terminals are directly connected to the central computer but not to one another.

61. In a **star** network configuration, the central computer acts as network controller.

62. **Star** network configurations are used in timesharing systems with the central node being the timesharing "host" computer.

Bus Networks

63. A **bus** network configuration is essentially a single multidrop line shared by many nodes.

64. On a **bus** network configuration, a message can be broadcast to all the nodes simultaneously.

65. **Bus** network configurations are more reliable than stars and rings because a node failure does not affect the other nodes.

Ring Networks

66. **Ring** network configurations are common network arrangements for local environments.

67. **Ring** network configurations do not have a central computer system controlling the operation of the entire system and use point-to-point lines.

68. Messages from one computer to another on a ring network configuration must be **addressed** to the receiver.

69. **Token passing** is commonly used in ring network configurations to determine which node transmits next.

70. A bit pattern called a **token** is passed from node to node until it is taken by a node that is ready to transmit.

71. Ring network configurations are formed with **point-to-point** lines.

Local Networks

72. **Local** networks are designed to handle limited geographic areas.

73. **Local** networks are generally company owned.

74. **Ethernet** is a local network developed by Xerox.

Case Study 7-1: The Ethernet Local Network

75. Ethernet uses a **coaxial** copper cable to connect the various pieces of information equipment.

76. Information on the **Ethernet** system travels over the cable in packets of data.

77. To decide which piece of equipment on the network gets to communicate next, Ethernet uses the **collision** concept.

78. The collision concept on Ethernet is also called **contention** .

79. The **collision** concept of Ethernet allows it to operate without central network control.

80. If an individual device fails on the Ethernet network, the network keeps functioning in what is called **distributed network** control.

81. One disadvantage of Ethernet is that it is a **baseband** rather than a broadband network.

82. **Private Branch Exchanges (PBXs)** , often called CBXs, can carry both voice and data communications so that local networks can be implemented by use of standard telephone lines.

Multiplexors

83. A **multiplexor** is a device that enables sharing of a single high-speed line.

84. A **multiplexor** divides the capacity of a high-speed line to make it appear to be several low-speed lines.

85. **Time-division** multiplexing dedicates the line to each terminal for brief periods of time.

86. **Frequency-division** multiplexing assigns different frequency ranges to different terminals.

87. Radio and TV stations transmit in the same manner as **frequency-division** multiplexing.

Statistical Multiplexors

88. A **statistical** multiplexor gives unused line capacity to other terminals which are ready to transmit.

89. A statistical multiplexor is also called a **stat mux**.

Concentrators

90. Airline transportation is similar to **concentrator** operation with data communications.

91. Unlike multiplexors, **concentrators** can "absorb" huge amounts of information and hold the data for long periods before transmitting it.

Common Carriers

92. **Common carriers** are companies authorized by the government to provide communications services to the public.

93. The **Carterphone** decision brought many more competitors into the data communications industry.

94. Teleconferencing, electronic mail, and packet switching are examples of **value-added** services.

95. The value-added service that allows executives to hold face-to-face meetings is called **teleconferencing** .

96. **Electronic mail** is a value-added service that allows messages to be sent over computer networks instead of through the postal service.

97. **Packet switching** is value-added service which provides error correction and detection capabilities through prefixed identifiers on blocks of data.

Expectations for the Future

98. Advancements in **fiber optics** are expected to revolutionize cable technology.

99. Silicon chips are expected to be replaced with chips made from a compound called **gallium arsenide** .

Chapter 8
Structured Programming

Multiple Choice

What Is a Computer Program?

1. Computer programs may be written in many programming languages but a computer can only understand its own ______.
   (a) pseudocode
   (b) input instructions
   (c) machine language *
   (d) arithmetic instructions
   (e) production program

2. The ______ instructions cause data to be read into the computer's primary storage.
   (a) output
   (b) arithmetic
   (c) comparison
   (d) data movement
   (e) input *

3. The instructions that cause data to be written from primary storage to a printer or secondary storage device are ______ instructions.
   (a) input
   (b) arithmetic
   (c) data movement
   (d) output *
   (e) file and record definition

4. ______ instructions cause operations such as addition and division.
   (a) Arithmetic *
   (b) Output
   (c) Data movement
   (d) Input
   (e) Control

5. The ______ instructions determine if numbers are equal, greater than, or less than one another.
   (a) control
   (b) comparison *
   (c) arithmetic
   (d) input

(e) data definition

6. ______ instructions specify the order in which instructions are to be performed.
(a) Comparison
(b) Arithmetic
(c) Input
(d) Control *
(e) Data movement

7. The instructions that cause information to be moved between locations in primary storage are ______ instructions.
(a) input
(b) output
(c) data movement *
(d) data definition
(e) file and record definition

8. ______ instructions specify the data to be used by a program.
(a) Data definition *
(b) Control
(c) Input
(d) File and record definition
(e) Comparison

9. The instructions that allow programmers to specify data to be processed by a program and which secondary storage devices contain the data are called ______ instructions.
(a) data definition
(b) file and record definition *
(c) control
(d) input
(e) arithmetic

The Program Development Process

10. The gradual process of developing a solution to a problem is called ______.
(a) pseudocode
(b) debugging
(c) flowcharting
(d) documentation
(e) top-down stepwise refinement *

11. The ______ process involves determining what is wrong with a program and making the necessary corrections.
(a) documentation
(b) top-down stepwise refinement

(c) debugging *
(d) flowcharting
(e) selection structure

12. The ______ helps people to understand a program and helps to facilitate modificaton.
(a) flowcharting
(b) pseudocode
(c) debugging
(d) documentation *
(e) top-down stepwise refinement

13. When a program is put into regular use it becomes a ______ program.
(a) production *
(b) structured
(c) egoless
(d) repetition
(e) sequence

The Program Design Phase

14. Programs should be ______, that is, they should run quickly and with minimial use of expensive resources.
(a) robust
(b) efficient *
(c) correct
(d) reliable
(e) maintainable

15. A program should be ______ in that it should operate properly under a wide range of legitimate inputs, and should be easy to modify.
(a) maintainable
(b) efficient
(c) reliable
(d) correct
(e) robust *

Flowcharting

16. Programs can be designed using different symbols in a process called ______.
(a) pseudocode
(b) documentation
(c) flowcharting *
(d) subscripts
(e) nested structures

17. The order in which operations are to be performed is indicated on a flowchart by ______.
(a) a decision symbol
(b) flowlines *
(c) an input/output symbol
(d) a terminal symbol
(e) a process symbol

18. The ______ symbol means a condition is to be tested.
(a) connector
(b) terminal
(c) input/output
(d) process
(e) decision *

19. The ______ symbol has two arrows leaving it, one to be followed if a condition is true and the other if the condition is false.
(a) decision *
(b) process
(c) input/output
(d) terminal
(e) connector

20. The ______ symbol indicates the beginning and the end of a program.
(a) connector
(b) input/output
(c) terminal *
(d) decision
(e) process

21. The comparison operator that means is not equal to is ______.
(a) >
(b) <> *
(c) <
(d) >=
(e) <=

22. The comparison operator that means is greater than is ______.
(a) <
(b) <=
(c) <>
(d) > *
(e) >=

23. The comparison operator that means is less than is ______.
(a) >=

(b) >
(c) < *
(d) <=
(e) <>

24. The comparison operator that means is greater than or equal to is ______.
(a) >= *
(b) <
(c) <=
(d) <>
(e) >

25. The comparison operator that means is less than or requal to is ______.
(a) <>
(b) >=
(c) <
(d) <= *
(e) >

Testing and Debugging

26. The removing of errors from a program is referred to as ______.
(a) flowcharting
(b) debugging *
(c) pseudocode
(d) spaghetti code
(e) indexing

27. A ______ error occurs when a programmer writes an invalid statement.
(a) run-time
(b) execution
(c) syntax *
(d) index
(e) logic

28. The most difficult errors to find are ______ errors.
(a) execution
(b) syntax
(c) index
(d) logic *
(e) run-time

Structured Programming

29. The ______ structure indicates that one operation is to be performed immediately after another.
   (a) control
   (b) repetition
   (c) selection
   (d) CASE
   (e) sequence *

30. The structure that chooses an alternative depending on whether a condition is true or false is the ______ structure.
   (a) CASE
   (b) control
   (c) selection *
   (d) repetition
   (e) sequence

31. The structure that indicates that certain operations are to be performed several times is the ______ structure.
   (a) IF-THEN
   (b) repetition *
   (c) CASE
   (d) control
   (e) selection

Pseudocode

32. ______ is also called structured English.
   (a) Pseudocode *
   (b) Flowcharting
   (c) Documentation
   (d) Structured walkthrough
   (e) IPO method

33. The English-like sentences used by programmers to explain what a program is to do are called ______.
   (a) IPO method
   (b) documentation
   (c) flowcharting
   (d) pseudocode *
   (e) structured walkthroughs

Other Control Structures

34. The structure that does nothing when a condition is false is the ______.

(a) CASE
(b) REPEAT-UNTIL
(c) IF-THEN *
(d) WHILE-DO
(e) ELSE

35. The ______ structure is normally used when a program must choose from more than two alternatives.
(a) ELSE
(b) WHILE-DO
(c) REPEAT-UNTIL
(d) IF-THEN
(e) CASE *

GOTO Elimination

36. The ______ statement causes a transfer of control to another part of the program.
(a) CASE
(b) GOTO *
(c) IPO
(d) array
(e) index

37. When a program flowchart has many intersecting flowlines it is referred to as ______.
(a) egoless programming
(b) transfer of control
(c) spaghetti code *
(d) structured walkthrough
(e) subroutine calls

Programming Psychology

38. In ______ all members of a project share their techniques and review each other's work.
(a) egoless programming *
(b) chief programmer team
(c) pseudocode
(d) top-down stepwise refinement
(e) structured programming

Structured Walkthroughs

39. With the technique of ______ a peer group review of an individual's work checks for accuracy.

(a) structured programming
(b) top-down stepwise refinement
(c) pseudocode
(d) structured walkthrough *
(e) chief programmer team

The Chief Programmer Team

40. When a system does not seem to hang together it is said to suffer from a lack of ______.
(a) spaghetti code
(b) syntax errors
(c) logic errors
(d) repetition structure
(e) conceptual integrity *

41. The ______ in the chief programmer team leads the team through the development of the new system.
(a) copilot
(b) editor
(c) chief programmer *
(d) tester
(e) program clerk

42. The person in the chief programmer team that can take over for the chief is the ______.
(a) language lawyer
(b) copilot *
(c) tester
(d) toolsmith
(e) program clerk

43. The ______ in the chief programmer team produces the documentation.
(a) administrator
(b) tester
(c) editor *
(d) secretary
(e) language lawyer

44. The person who handles all inputs and outputs in the chief programmer team is the ______.
(a) program clerk *
(b) language lawyer
(c) administrator
(d) tester
(e) copilot

45. The ______ constructs special programs for the chief in the chief programmer team.

(a) language lawyer
(b) copilot
(c) program clerk
(d) toolsmith *
(e) tester

46. The special programs that the toolsmith constructs are called ______.
(a) subscripts
(b) arrays
(c) indexes
(d) modules
(e) utilities *

47. The person in the chief programmer team that prepares special cases that check if the programs are running correctly is the ______.
(a) toolsmith
(b) tester *
(c) copilot
(d) administrator
(e) program clerk

48. The member of the chief programmer team who is an expert in the structured programming language being used is the ______.
(a) chief programmer
(b) program clerk
(c) language lawyer *
(d) tester
(e) editor

Case Studies in Structured Program Development

49. The process that shows sample input data, structured flowcharts and their corresponding pseudocode, and the outputs produced is the ______ method.
(a) IPO *
(b) GOTO
(c) CASE
(d) IF-THEN
(d) REPEAT-UNTIL

50. When a selection structure is embedded within an outer structure it is said to be ______.
(a) initialized
(b) predefined
(c) table searching
(d) nested *
(e) single-entry, single-exit structure

51. A special marker that indicates that a program has reached the end of data entry is called a ______.
   (a) array
   (b) sentinel value *
   (c) index
   (d) module
   (e) control break

52. When a variable is set to zero it is said to be ______.
   (a) formatted
   (b) moduled
   (c) initialized *
   (d) indexed
   (e) subscripted

53. The flowchart symbol that is a rectangle with vertical bars near its left and right borders is called a ______ symbol.
   (a) input/output
   (b) process
   (c) decision
   (d) terminal
   (e) predefined process *

54. ______ help in writing programs in modular fashion as a collection of program pieces.
   (a) Subroutines *
   (b) Indexes
   (c) Arrays
   (d) Tables
   (e) Subscripts

55. A program piece is also called a ______.
   (a) Index
   (b) array
   (c) subscript
   (d) module *
   (e) table

56. The last statement executed in any subroutine is the ______ statement.
   (a) Sentinel
   (b) Input
   (c) Return *
   (d) Data
   (e) End

57. Tables are often called ______.

(a) sentinels
(b) arrays *
(c) modules
(d) indexes
(e) entry points

58. A ______ is a list of related data entries.
(a) table *
(b) sentinel
(c) module
(d) subroutine
(e) index

59. Each entry in a table is referred to by using a(n) ______.
(a) array
(b) sentinel
(c) index *
(d) subroutine
(e) module

60. An index is also referred to as a(n) ______.
(a) module
(b) subroutine
(c) array
(d) subscript *
(e) sentinel

61. When a program switches from processing one group of records with one employee number to processing another group of records with another employee number it is using ______.
(a) sentinels
(b) control breaks *
(c) modules
(d) subscripts
(e) arrays

True/False

What Is a Computer Program?

1. A computer program is nothing more than a listing of the steps, or operations, to be performed to solve a problem. (F p186)

2. Computera are considered to be general-purpose instruments. (T p186)

3. Programs must be executed before they may be loaded. (F p186)

4. Most programs are written in machine language. (F p186)

5. Input instructions cause data to be transferred into a computer's primary storage. (T p186)

6. Arithmetic instructions can compare pairs of numbers to determine if they are equal. (F p186)

7. Control instructions specify the order in which instructions are to be performed. (T p187)

The Program Development Process

8. Specifications are prepared to describe precisely and clearly the problem to be solved. (T p187)

9. Determining what is wrong with a program and making the appropriate corrections is called testing. (F p187)

10. A program that has been put into regular use is called a production program. (T p187)

The Program Design Phase

11. A program that meets its specifications is said to be robust. (F p187)

12. A program that operates properly under a wide range of legitimate inputs is said to be robust. (T p187)

13. A program that continues to operate under unforeseen circumstances is said to be reliable. (T p187)

Flowcharting

14. In a flowchart, flowlines indicate how data is passed between symbols. (F p187)

15. In a flowchart, the shapes of the symbols indicate the nature of the operations to be performed. (T p187)

16. The shapes of the common flowcharting symbols have been standardized by an organization called ASCII. (F p187)

17. The oval-shaped symbol is called the process symbol. (F p188)

18. The diamond-shaped symbol indicates that a decision is to be made. (T p188)

19. The rectangle-shaped symbol indicates that an input/output operation is to be performed. (F p188)

20. The terminal symbol indicates that two widely separated places of a flowchart directly connect from one to the other. (F p188)

21. The decision symbol has one flowline in and two flowlines emerging from it. (T p188)

22. If a symbol has four straight sides, then that symbol always performs a calculation. (F p188)

23. A condition is a comparison statement that is true. (F p188)

24. On most computers, the six common comparison operators require two keystrokes each. (F p189)

25. The comparison operator "<>" means is less than and greater than. (F p189)

26. One disadvantage of flowcharts is that programmers attempting to solve the same problem may produce dramatically different flowcharts. (T p189)

Testing and Debugging

27. Bugs that appear before execution are normally syntax errors. (T p189)

28. The syntax of a programming language is the set of rules that governs the writing of valid statements in that language. (T p189)

29. Normally, misspelling a word of a programming language will cause a run-time error. (F p189)

30. Invalid or insufficient data read at execution time cause syntax errors. (F p189)

31. Division by zero is an example of a run-time error. (T p189)

32. Logic errors are serious errors that cause a program to terminate. (F p190)

33. An example of a logic error is a mailing list program that prints addresses without zip codes. (T p190)

34. Output controllers are people who try to spot errors in outputs as early as possible. (T p190)

35. Ideally, programs should be made bug free by extensive testing with a carefully designed set

of test cases. (F p190)

Structured Programming

36. Bohm and Jacopini proved that the most complex program logic could be expressed by the use of only the selection and repetition control structures. (F p190)

37. The sequence structure indicates that one operation is to be performed immediately after another. (T p191)

38. The selection structure and the repetition structure each depend upon specified conditions being true or false. (T p191)

39. All flowchart symbols have a single entry and a single exit. (F p191)

40. All structured programming control structures have a single entry and a single exit. (T p191)

Pseudocode

41. One reason for the popularity of pseudocode is that the text can be maintained by a computerized text editing system. (T p191)

42. The IF-THEN structure is a modified repetition structure that simply does nothing when the condition is false. (F p192)

43. The REPEAT-UNTIL structure is a variation of the WHILE-DO structure in which the test to determine if looping should continue is made after, rather than before, the action is performed. (T p192)

44. The REPEAT-UNTIL structure should be used when it is known in advance that the action may not have to be performed at all. (F p192)

45. The WHILE-DO must always be used when it is known in advance that the action will be performed at least once. (F p192)

46. The CASE structure is useful in selection situations in which the program must choose from more than two alternatives. (T p193)

GOTO Elimination

47. In the sequence structure, control automatically transfers to the next instruction.

(T p193)

48. The GOTO statement simply causes a transfer of control to a designated place in the program. (T p194)

49. The GOTO statement results in well-organized, easy to read, and easy to debug programs. (F 194)

50. Dijkstra argued that the GOTO statement leads to "spaghetti code." (T p194)

51. In some special cases even well-structured programs should use the GOTO. (T p194)

Programming Psychology

52. Gerald Weinberg proposed the notion of egoless programming. (T p195)

53. Dijkstra wrote *The Psychology of Computer Programming*. (F p195)

Structured Walkthroughs

54. Weinberg's concept of egoless programming led to the chief programmer team concept. (F p195)

55. A structured walkthrough is a peer group review of an individual's work. (T p195)

56. Structured walkthroughs are used universally in large organizations. (F p195)

The Chief Programmer Team

57. One advantage of team programming is that the designs that emerge have conceptual integrity (unity of design). (F p195)

58. Fred Brooks authored *The Mythical Man Month*. (T p196)

59. The structured walkthrough has been likened to the operation of a surgical team. (F p196)

60. The chief programmer must be a "super programmer." (T p196)

61. The administrator in a chief programmer team serves as the alter ego of the chief. (F p196)

62. The copilot in a chief programmer team is able to take over in the chief's absence. (T p196)

63. In a chief programmer team, the program clerk handles all inputs, outputs, program files, backup files, and the like. (T p196)

64. In a chief programmer team, the tester constructs special programs called utilities that support the chief's efforts. (F p196)

65. In a chief programmer team, the language lawyer is an expert in the structured programming language being used in the project. (T p196)

66. The most serious problem with the chief programmer team concept is that chief programmers are hard to find. (T p196)

Case Studies in Structured Program Development

67. The IPO method stands for "input-processing-output." (T p197)

68. Top-down stepwise refinement is a gradual process used to develop structured programs and systems of programs. (T p197)

69. A structure embedded within another structure is said to be nested within the outer structure. (T p200)

70. The DO portion of the WHILE-DO structure may contain several statements. (T p200)

71. The REPEAT-UNTIL structure first tests to see if looping should continue, and if it should the body of the loop is executed. (F p201)

72. The WHILE-DO structure continues looping as long as the test condition remains true. (T p201)

73. The REPEAT-UNTIL continues looping as long as the condition remains false. (T p201)

74. The condition one would use in the REPEAT-UNTIL decision symbol is the opposite of the condition one would use in the corresponding WHILE-DO decision symbol. (T p201)

75. A sentinel-value should be used for controlling a loop when the number of repetitions is not known in advance. (T p203)

76. The sentinel value -1 is appropriate for a program that calculates average winter temperatures. (F p203)

77. The process of setting a variable to its beginning value is called initializing the variable. (T p206)

78. The predefined process symbol is used in a flowchart to indicate that a subroutine is to be called. (T p206)

79. Subroutines are examples of program pieces called modules. (T p207)

80. The Return statement is the last statement executed in any subroutine. (T p208)

81. The Return statement causes the main program to continue executing at the first statement before the statement that called the subroutine. (F p208)

82. An input variable is one that is supplied to a subroutine by the main program. (T p208)

83. An output variable is one with a value that is set by the subroutine and passed back to the main program. (T p208)

84. Tables are called arrays in computerese. (T p208)

85. Another name for index is array. (F p208)

86. A control break occurs when a program determines that a certain key piece of information on a current record matches a key piece of information on the previous record. (F p211)

Fill-in-the-Blanks

What Is a Computer Program?

1. A **computer program** is a procedure followed by a computer to solve problems.

2. A computer program is a listing of the steps, or **operations** to be performed and an indication of the order in which they are to be performed.

3. Computer programs are written in various **programming languages** .

4. A particular computer can only understand its own **machine** language.

5. An **input** instruction causes data to be read into the computer's primary storage from such devices as a keyboard or disk.

6. An **output** instruction causes data to be written from the computer's primary storage to such devices as a printer or disk.

7. An **arithmetic** instruction causes the computer to perform such operations as addition, subtraction, multiplication, and division.

8. A **comparison** instruction causes the computer to compare pairs of numbers to determine if one number is equal to, greater than, or less than the other.

9. A **control** instruction specifies the order in which instructions are to be performed.

10. A **data movement** instruction causes data items to be moved between locations in primary storage.

11. A **data definition** instruction specifies the data to be used by a program.

12. **File and record definition** instructions allow the programmer to specify which files of data will be processed, which secondary storage device contains the files, and what the characteristics of these files are.

The Program Development Process

13. Preparing **specifications** that describe precisely and clearly the problem to be solved is one step to writing good programs.

14. The gradual solution to solving a problem using design aids such as flowcharts and pseudocode is called the **top-down stepwise refinement** process.

15. **Debugging** involves determining what is wrong with a program and making the appropriate corrections.

16. **Documentation** is carried on throughout all phases of program development.

17. When a program is put into regular use it becomes a **production** program.

The Program Design Phase

18. A program should be **efficient** , that is, it should execute quickly with minimal use of expensive resources such as primary storage.

19. A program should be **robust** in that it operates properly under a wide range of legitimate inputs, and it should be easy to modify.

20. A program should be **reliable** in that it continues to operate under unforeseen circumstances.

Flowcharting

21. A method of designing programs using symbols is called **flowcharting** .

22. The symbols in a flowchart are connected by **flowlines** ending in arrows.

23. **Flowlines** show the order of operations in a flowchart.

24. The symbols used in flowcharting are punched out of on a plastic flowcharting **template** .

25. A **condition** is simply a comparison statement.

26. The **terminal** symbol in a flowchart indicates the beginning or end of a program.

27. A **process** symbol in a flowchart indicates that a calculation should be performed.

28. The **decision** symbol in a flowchart indicates that a condition is to be tested.

Testing and Debugging

29. The computerese term for an error is **bug**.

30. A **syntax** error occurs when a programmer writes an invalid statement.

31. Run-time errors, or **execution** errors, occur while a program is running.

32. A **logic** error does not always cause a program to terminate.

33. Output **controllers** are people who help spot problems such as logic errors.

34. The comparison operator that means "is equal to" is =.

35. <> is the comparison operator that means "is not equal to."

36. The comparison operator that means "is greater than" is >.

37. The "is less than" comparison operator is <.

38. >= is the comparison operator that means "is greater than or equal to."

39. <= is the comparison operator that means "is less than or equal to."

Structured Programming

40. Most complex program logic can be expressed by the use of three **control structures** ; sequence, selection, and repetition.

41. The **sequence** structure indicates that one operation is to be performed immediately after another.

42. The **selection** structure designates that the operations to be performed in the future depend on whether a certain condition is true or false.

43. The **repetition** structure indicates that certain operations are to be repeated while some condition remains true.

44. The use of control structures is the key to a programming methodology called **structured programming** .

Pseudocode

45. **Pseudocode** uses English-like sentences to write an explanation of what a program is supposed to do.

46. Another name for **pseudocode** is structured English.

Other Control Structures

47. The **IF-THEN** structure is a modified selection structure that simply does nothing when the condition is false.

48. The **REPEAT-UNTIL** structure is a variation of the WHILE-DO structure in which the the test to determine if looping should continue is made after the action is performed.

49. The **CASE** structure is useful in selection situations in which the program must choose from more than two alternatives.

GOTO Elimination

50. Built into each of the control structures is the use of the **transfer of control** , the ability of a program to resume execution at some point in the program possibly other than the next instruction.

51. A **GOTO** statement causes a transfer of control to another part of a computer program.

Programming Psychology

52. Gerald Weinberg proposed the notion of **egoless** programming.

53. **Egoless programming** encourages all members of a project to share their techniques and to review one another's work.

Structured Walkthroughs

54. A **structured walkthrough** is a peer group review of an individual's work.

The Chief Programmer Team

55. Designs that emerge from a team can lack **conceptual integrity** ; the system just doesn't seem to hang together.

56. In the **chief programmer team** , all other members support the chief.

57. In the chief programmer team, the **chief programmer** leads the team.

58. The alter ego to the chief in the chief programmer team is the **copilot** .

59. The member of the chief programmer team who handles all administrative matters is the **administrator** .

60. The job of the **editor** in the chief programmer team is proofreading and correcting the documentation.

61. The **program clerk** in the chief programmer team handles all inputs, outputs, program files, backup files, and so on.

62. The member of the chief programmer team who constructs special programs to support the chief's efforts is the **toolsmith** .

63. The Toolsmith in a chief programmer team builds programs called **utilities** or software tools.

64. The member of the chief programmer team who prepares test cases to ensure programs are running properly is the **tester** .

65. The **language lawyer** is an expert in the structured programming language being used by the chief programmer team.

Case Studies in Structured Program Development

66. In the **IPO method** of structured program development, a sample of input data is shown,

a structured flowchart and the corresponding structured pseudocode are presented to show what processing is to be performed, and the outputs that would be produced from the processing of the sample inputs are given.

Case Study 8-1: Simple Decision Making

67. A **selection** structure is used to determine which of two numbers is larger.

Case Study 8-2: Complex Decision Making

68. When a second structure is embedded within another structure it is called a **nested** structure.

Case Study 8-3: Counter-Controlled Looping with the WHILE-DO Structure

69. A **counter** can be used to control a loop only when it is known in advance how many repetitions of the loop are required.

Case Study 8-4: Counter-Controlled Looping with the REPEAT-UNTIL Structure

70. The **REPEAT-UNTIL** structure continues looping as long as the condition remains false.

71. The **WHILE-DO** continues looping as long as the test condition remains true.

Case Study 8-5: Counting, Totaling, and Averaging with a Sentinel-Controlled Loop

72. A **sentinel value** is a special marker used to indicate the end of data being entered.

Case Study 8-7: Piecework Payroll for One Employee: Introducing Subroutines

73. When a variable is **initialized** its beginning value is set to zero (or some other meaningful value).

74. The **predefined process symbol** in a flowchart is a rectangle with two vertical bars near its left and right borders.

75. **Subroutines** help in writing programs as a collection of program pieces.

76. When designing a new program, each program piece is called a **module**.

Case Study 8-8: Piecework Payroll: Subroutines and Table Searching

77. The last statement executed in any subroutine is the **Return** statement.

78. An **input** variable is one that is supplied to the subroutine by the main program.

79. Tables are called **arrays** in computerese

80. The process of looking through a table to find data is called **table searching** .

81. Each entry in a table is referred to by using an index or **subscript** .

Case Study 8-9: Piecework Payroll for Several Employees: Control Break Processing

82. When a computer is processing many groups of related records, when the start of a new group is encountered, this is called a **control break** .

Chapter 9
Programming Languages

Multiple Choice

Programming Languages

Machine Language

1. Each computer can understand only one language--its ______ language.
   (a) programming
   (b) machine *
   (c) macro
   (d) universal programming
   (e) assembly

2. Because machine languages are so closely related to the structure of a particular computer, they are said to be ______.
   (a) procedure-oriented
   (b) problem-oriented
   (c) machine dependent *
   (d) symbol manipulation
   (e) sources

Assembly Language

3. The term ______ is generally associated with assembly languages.
   (a) mnemonic *
   (b) quick-and-dirty
   (c) procedure-oriented
   (d) problem-oriented
   (e) optimizing

4. ______ are English-like abbreviations used for many machine language instructions.
   (a) p-Code
   (b) CODASYL
   (c) UPL
   (d) Mnemonics *
   (e) Compilers

5. Mnemonics are translated into machine language by ______.

(a) interpreters
(b) artificial intelligence
(c) symbol manipulation
(d) p-Code
(e) assemblers *

6. The program of instructions written in assembly language is known as the ______ program.
(a) object
(b) translator
(c) source *
(d) generator
(e) procedure

7. An assembler program translates a source program into a(n) ______ program.
(a) procedure
(b) object *
(c) generator
(d) source
(e) translator

Macro Instructions

8. A(n) ______ is translated into several machine language instructions.
(a) mnemonic
(b) object instruction
(c) interpreter
(d) compiler
(e) macro instruction *

9. An assembler program must perform a(n) ______ on a macro instruction.
(a) macro expansion *
(b) production
(c) mnemonic
(d) interpretation
(e) hierarchical data structuring

High-Level Languages

10. High-level language programs are translated by ______ into machine language.
(a) ANSI
(b) macro instructions
(c) compilers *
(d) interpreters
(e) mnemonics

11. A(n) ______ can run a source program directly.
    (a) translator program
    (b) compiler
    (c) mnemonic
    (d) interpreter *
    (e) macro expansion

Procedure-Oriented and Problem-Oriented Languages

12. ______ languages are general purpose and can be used to solve many types of problems.
    (a) Program generator
    (b) Problem-oriented
    (c) Object
    (d) Assembly
    (e) Procedure-oriented *

13. ______ languages are designed to solve specific types of problems.
    (a) Assembly
    (b) Procedure-oriented *
    (c) Object
    (d) Problem-oriented
    (e) Program generator

"Quick-and-Dirty" Compilers and Optimizing Compilers

14. In ______ environments, programmers are more concerned with debugging programs than having them run efficiently.
    (a) translation
    (b) program generator
    (c) program development *
    (d) specifications
    (e) production

15. ______ translate programs fast but the machine language programs produced may use more primary storage than they really need.
    (a) "Quick-and-dirty" compilers *
    (b) Interpreters
    (c) Assemblers
    (d) Optimizing compilers
    (e) Mnemonics

16. In ______ environments programs, must run efficiently.
    (a) program development

(b) expansion
(c) interpreter
(d) production *
(e) optimizing

17. ______ usually take more time to translate a program but the machine language programs produced run faster and use less primary storage.
(a) Interpreters
(b) "Quick-and-dirty" compilers
(c) Macro instructions
(d) Mnemonics
(e) Optimizing compilers *

A Survey of High-Level Languages

BASIC

18. The ______ high-level language was designed to be simple to learn and easy to use.
(a) COBOL
(b) BASIC *
(c) PL/1
(d) ALGOL
(e) FORTRAN

19. The ______ language was developed by John Kemeny and Thomas Kurtz.
(a) PL/1
(b) RPG
(c) FORTRAN
(d) BASIC *
(e) APL

20. ______ was originally developed for interactive use on timesharing computers.
(a) BASIC *
(b) COBOL
(c) LISP
(d) Forth
(e) Ada

21. In the ______ programming language, the editing capabilities are built right into the language.
(a) ALGOL
(b) FORTRAN
(c) BASIC *
(d) RPG
(e) Pl/1

FORTRAN

22. The oldest high-level language still in use today is ______.
    (a) BASIC
    (b) PL/1
    (c) LISP
    (d) LOGO
    (e) FORTRAN *

23. The first programming language to become standardized was ______.
    (a) LOGO
    (b) FORTRAN *
    (c) BASIC
    (d) COBOL
    (e) Pascal

COBOL

24. The ______ programming language was developed for business data-processing applications that handle large amounts of data, and edit it for presentation.
    (a) COBOL *
    (b) LOGO
    (c) LISP
    (d) Pascal
    (e) Ada

25. ______ developed COBOL; its participants included Grace Hopper.
    (a) ANSI
    (b) UPL
    (c) CODASYL *
    (d) VHLL
    (e) U.S. Defense Department

26. ______ language programs use English words, contain paragraphs and sentences, and use punctuation.
    (a) Pascal
    (b) Ada
    (c) BASIC
    (d) COBOL *
    (e) Forth

27. Every ______ program is broken up into four major divisions.
    (a) Forth

(b) LOGO
(c) LISP
(d) Ada
(e) COBOL *

28. The ______ division of a COBOL program includes the author's name and the date the program was compiled.
(a) file
(b) identification *
(c) configuration
(d) data
(e) environment

29. The ______ division of a COBOL program defines the computer on which the program is to be run and the files to be processed.
(a) environment *
(b) data
(c) identification
(d) file
(e) input-output

30. The ______ division of a COBOL program specifies the format of each data item used by the program.
(a) input-output
(b) file
(c) data *
(d) identification
(e) environment

31. The ______ division of a COBOL program specifies the actual calculations to be performed.
(a) ennvironment
(b) identification
(c) file
(d) procedure *
(e) input-output

32. COBOL tends to be ______ requiring fewer additional comments and remarks than assembly language programs.
(a) multitasking
(b) symbol manipulation
(c) program generator
(d) specification
(e) self-documenting *

33. The feature of COBOL that allows for increasing levels of detail is called ______.
(a) hierarchical data structuring *

(b) multitasking
(c) program generation
(d) symbol manipulation
(e) specification

PL/1

34. The structured programming language designed to contain the best features of FORTRAN and COBOL is ______.
(a) BASIC
(b) PL/1 *
(c) Ada
(d) LISP
(e) Forth

35. The ______ programming language was introduced by IBM along with its System/360 series of computers.
(a) Forth
(b) APL
(c) PL/1 *
(d) RPG
(e) C

RPG

36. The ______ language requires the programmer to fill out special forms that tell the computer what is to be done.
(a) PL/1
(b) Query
(c) C
(d) RPG *
(e) ALGOL

37. Programmers using RPG fill in ______ forms.
(a) data division
(b) macro instruction
(c) self-documenting
(d) program generator
(e) specifications *

38. Since the RPG translator writes the actual program, the language is considered a ______.
(a) program generator *
(b) symbol manipulation language
(c) universal programming language

(d) software package
(e) very-high-level language

39. The ______ programming language was developed to make it easier to produce report-writing programs.
(a) BASIC
(b) RPG *
(c) LOGO
(d) LISP
(e) C

APL

40. ______ is a highly mathematical language developed by Kenn Iverson.
(a) RPG
(b) Pascal
(c) C
(d) APL *
(e) LOGO

41. The programming language that has its own character set is ______.
(a) LOGO
(b) C
(c) APL *
(d) LISP
(e) Pascal

42. The ______ programming language requires a special keyboard.
(a) UCSD Pascal
(b) C
(c) LISP
(d) APL *
(e) LOGO

Pascal

43. The programming language developed by Nicklaus Wirth is ______.
(a) LOGO
(b) Pascal *
(c) COBOL
(d) C
(e) LISP

44. The programming language that is a descendant of ALGOL is ______.

(a) Pascal *
(b) Forth
(c) COBOL
(d) LISP
(e) BASIC

45. Pascal compilers produce instructions in ______.
(a) UCSD Pascal
(b) symbol manipulation language
(c) p-Code *
(d) UCSD p-System
(e) universal programming language

46. ______ is a version of Pascal that is a complete programming environment containing its own operating system.
(a) Query
(b) UPL
(c) VHLL
(d) UCSD p-System *
(e) UNIX

47. ______ is one of the best programming languages available for expressing algorithms.
(a) BASIC
(b) C
(c) Pascal *
(d) LOGO
(e) UNIX

C

48. The ______ language was used to code the UNIX operating system.
(a) C *
(b) LOGO
(c) LISP
(d) Query
(e) Ada

49. The ______ programming language was developed at Bell Laboratories.
(a) Ada
(b) LISP
(c) LOGO
(d) C *
(e) Forth

Ada

50. The ______ programming language allows for multitasking.
    (a) C
    (b) Forth
    (c) BASIC
    (d) Pascal
    (e) Ada *

51. The base for the ______ programming language is Pascal.
    (a) BASIC
    (b) Ada *
    (c) C
    (d) LOGO
    (e) Forth

52. The development of the ______ programming language was sponsored by the U.S. Defense Department.
    (a) Forth
    (b) C
    (c) LOGO
    (d) Ada *
    (e) LISP

Forth

53. The ______ programming language was developed by Charles More at the Kitts Peak Observatorry.
    (a) LISP
    (b) C
    (c) LOGO
    (d) Pascal
    (e) Forth *

54. The ______ programming language automatically saves all of its functions building up a rich set of capabilities.
    (a) Forth *
    (b) LOGO
    (c) COBOL
    (d) C
    (e) BASIC

LISP

55. ______ was one of the first symbol manipulation languages.
(a) LOGO
(b) C
(c) LISP *
(d) Forth
(e) FORTRAN

56. The ______ language was developed by John McCarthy at MIT.
(a) Forth
(b) LISP *
(c) LOGO
(d) Pascal
(e) Ada

57. One of the first languages used for text processing was ______.
(a) LOGO
(b) C
(c) Ada
(d) LISP *
(e) BASIC

58. The programming language preferred in the field of artificial intelligence is ______.
(a) LISP *
(b) BASIC
(c) LOGO
(d) C
(e) Pascal

Query Languages

59. ______ languages allow computer users to request information from computers by typing simple English sentences.
(a) Universal programming
(b) Very-high-level
(c) Problem-oriented
(d) Query *
(e) Procedure-oriented

60. Many people believe that ______ languages are the key to making computers more user-friendly.
(a) very-high-level
(b) universal programming
(c) problem-oriented
(d) procedure-oriented
(e) query *

A Universal Programming Language

61. A disadvantage to ______ is that a huge amount of software would have to be replaced.
   (a) ANSI
   (b) UNIX
   (c) UPL *
   (d) VHLL
   (e) p-Code

62. With ______ only one programming language would have to be learned.
   (a) p-Code
   (b) UPL *
   (c) UNIX
   (d) VHLL
   (e) ANSI

Software Packages

63. By purchasing ______, users can obtain needed applications capabilities immediately.
   (a) turtle graphics
   (b) macro instructions
   (c) interpreters
   (d) software packages *
   (e) procedure-oriented languages

64. Most personal computer users purchase ______ rather than developing their own programs.
   (a) software packages *
   (b) procedure-oriented languages
   (c) interpreters
   (d) turtle graphics
   (e) macro instructions

Trends

65. The use of general-purpose ______ will allow users to write programs with fewer statements.
   (a) machine languages
   (b) macro instructions
   (c) VHLLs *
   (d) ANSI
   (e) interpreters

66. With the use of ______ programming languages, programmers will be able to specify that many tasks are to be performed simultaneously.
(a) machine language
(b) concurrent *
(c) query
(d) macro
(e) artificial intelligence

A Brief Introduction to LOGO

67. ______ was developed by Seymour Papert of MIT.
(a) LISP
(b) Pascal
(c) C
(d) RPG
(e) LOGO *

68. The ______ programming language was developed to facilitate the educational process especially with young children.
(a) BASIC
(b) COBOL
(c) Pascal
(d) LOGO *
(e) LISP

69. The powerful graphics capabilities of the LOGO programming language are called ______ graphics.
(a) mnemonics
(b) macro
(c) turtle *
(d) rabbit
(e) symbol

True/False

Machine Language

1. Each computer can directly understand only its machine language. (T p220)

2. Machine languages are considered to be machine dependent. (T p220)

3. Programs written in machine language are considered to be portable. (F p220)

Assembly Language

4. Today programmers rarely write programs in high-level languages. Instead, they use clearer machine languages and assembly languages. (F p222)

5. Machine language programs use mnemonics in their programs. (F p222)

6. An assembler is a translator program. (T p222)

7. A translator program generally converts an object program to a source program. (F p222)

8. Programs can be written faster with assembly language than with machine language. (T p223)

9. Assembly language programs are machine dependent and portable. (F p223)

10. Assembly language programs appear to be in computerese. (T p223)

11. The major use of assembly languages today is in writing operating systems. (T p223)

Macro Instructions

12. A single machine language instruction is generally translated into several macro instructions. (F p223)

13. Whenever an assembler program encounters a macro instruction it performs a macro expansion. (T p224)

14. The macro SUM A,B,C might produce LOAD A, ADD B, STORE C. (T p224)

High-Level Languages

15. Assembly language programming statements look like everyday English and use familiar mathematical notations. (F p224)

16. The translator programs that convert high-level language programs to machine language programs are called compilers. (T p224)

17. An interpreter is a translator program that runs a source program directly. (T p225)

18. An interpreter produces a machine language program. (F p225)

19. On every run of an interpreted program, the interpreter has to determine exactly what

every statement in the program means. (F p225)

20. As faster processors become more economical, interpreters will tend to be used less. (F p225)

Procedure-Oriented and Problem-Oriented Languages

21. Procedure-oriented languages are generally considered to be special-purpose languages. (F p225)

22. Most of the popular programming languages are considered to be procedure oriented. (T p225)

23. Problem-oriented languages are designed to solve specific types of problems. (T p225)

24. Procedure-oriented languages generally allow programmers to solve problems with fewer statements than are needed for problem-oriented languages. (F p225)

25. RPG is generally considered to be a problem-oriented language. (T p225)

"Quick-and-Dirty" Compilers and Optimizing Compilers

26. In program development environments, programmers are more concerned with debugging programs quickly than with having them run efficiently. (T p225)

27. When a program is being debugged, "quick-and-dirty" compilers are used. (T p225)

28. Once a program has been debugged, it is put into production where it is run regularly to accomplish useful work. (T p226)

29. In production environments, optimizing compilers are used. (T p226)

BASIC

30. BASIC was developed at IBM by John Kemeny and Thomas Kurtz. (F p226)

31. BASIC was originally developed for use on batch processing computers. (F p226)

32. BASIC is the most widely used programming language in the world. (T p226)

33. Most popular languages other than BASIC generally have only interpretive versions. (F p226)

34. BASIC programs are always processed in increasing statement number order. (F p226)

35. BASIC is an ideal programming language for writing large systems. (F p227)

36. In BASIC, the program editing capabilities are generally built right into the language. (T p227)

37. It is likely that a standardized version of BASIC will emerge soon. (F p227)

38. BASIC is generally considered to be a structured programming language. (F p227)

FORTRAN

39. FORTRAN was the first widely-used high-level language. (T p227)

40. FORTRAN is the oldest high-level language still in regular use. (T p227)

41. FORTRAN was developed at Dartmouth College in the mid-1950s. (F p227)

42. FORTRAN is oriented to performing complex mathematical computations. (T p227)

43. FORTRAN has never been widely used by businesses for data processing because of the availability of better languages. (F p227)

44. IBM considered FORTRAN to be an important corporate product and restricted its use outside the company. (F p227)

45. FORTRAN was the first progrmming language to undergo industry standardization. (T p227)

COBOL

46. Two characteristics of business data processing applications are efficient handling of large volumes of data and editing data for neat presentation. (T p227)

47. COBOL was developed by IBM in 1959. (F p227)

48. Grace Hopper pioneered the development of Ada. (F p227)

49. COBOL is generally considered to be more user-friendly than FORTRAN. (T p227)

50. Many people complain that COBOL programs are too wordy. (T p227)

51. Every COBOL program is broken up into five major divisions. (F p229)

52. The Identification Division in COBOL includes information such as the author's name and the date the program was compiled. (T p229)

53. The Configuration Section of a COBOL program appears in the Indentification Division. (F p229)

54. The Input-Output Section of a COBOL program specifies the files to be processed and the auxiliary storage devices that contain those files. (T p229)

55. The File Section of a COBOL program shows how data is organized in the records for each file. (T p229)

56. The Working-Storage Section of a COBOL program defines the variables and constants used in the program. (T p229)

57. The Procedure Division of a COBOL program specifies the actual calculations to be performed. (T p229)

58. Assembly language programs tend to be self-documenting. (F p229)

59. Long variable names are discouraged in COBOL programming. (F p229)

60. The MERGE verb in COBOL is used to combine two sorted files to form a single file. (T p229)

61. The looping statement in COBOL is PERFORM. (T p229)

62. The ACCEPT statement in COBOL allows a new user onto the timesharing system. (F p229)

63. The CALL statement in COBOL is used to communicate with another computer in a data communications network. (F p229)

64. The File Section in COBOL appears in the Data Division. (T p230)

65. The Input-Output Section in COBOL appears in the Environment Division. (T p230)

66. The "X" picture character in COBOL means "any character." (T p230)

67. The "9" picture character in COBOL means "any digit." (T p230)

68. In COBOL, PIC 9(7) is a reasonable picture for a seven-digit telephone number. (T p230)

69. COBOL is popular among personal computer users. (F p230)

PL/1

70. PL/1 is a structured programming language that contains the better features of both FORTRAN and COBOL. (T p230)

71. PL/1 appeared with IBM's System/360 series of computers and replaced FORTRAN and COBOL. (F p230)

RPG

72. RPG is a problem-oriented language developed by IBM in the 1960s. (T p234)

73. RPG is actually a program generator rather than a simple compiler. (T p234)

74. One key capability provided by RPG II is the ability to manipulate information stored in a database. (F p235)

APL

75. One problem with APL is that it has its own unique character set. (T p235)

76. APL is a highly mathematical language requiring the use of many special symbols. (T p235)

Pascal

77. The programming language Pascal was developed by Kenn Iverson at IBM. (F p235)

78. The Pascal programming language is a descendant of COBOL. (F p235)

79. One problem with the Pascal programming language is that it is a large language that tries to be all things to all people. (F 235)

80. Pascal is a general-purpose, problem-oriented language. (F p235)

81. The following Pascal code is an example of a defined data type:
    DAYOFTHEWEEK=(MON,TUE,WED,THU,FRI,SAT,SUN) (T p235)

82. Pascal compilers generally translate Pascal statements into machine language. (F p235)

83. PL/1 programs are generally translated into an intermediate language called p-Code. (F p235)

84. One popular version of Pascal is the p-System developed at the University of Californis at San Diego. (T p236)

C

85. The C programming language was developed at IBM and used to code the UNIX operating system. (F p236)

86. One particularly interesting feature of C is that it has aspects of both assembly and high-level languages. (T p236)

87. The use of C is increasing rapidly, particularly among UNIX users. (T p236)

Ada

88. PL/1's influence is apparent throughout the final version of Ada. (F p236)

89. Ada allows a programmer to specify that many tasks are to be performed simultaneously. (T p236)

Forth

90. The Forth programmer defines functions in terms of already defined functions, building up an ever-richer set of capabilities. (T p236)

91. Each time the programmer uses Forth, all previously defined functions become available. (T p236)

LISP

92. LISP programs can appear saturated with parentheses. (T p236)

93. LOGO was one of the first symbol manipulation languages. (F p236)

94. LISP has become the preferred language of artificial intelligence. (T p238)

95. (NULL X) in LISP tests if the list X is empty. (T p238)

96. (CDR X) in LISP indicates the first element of list X. (F p238)

Query Languages

97. Query languages allow computer users to request information from computers by typing simple English sentences. (T p 238)

A Universal Programming Language

98. All trends point to the eventual development of a Universal Programming Language. (F p238)

Software Packages

99. Most software used on personal computers and small business computers is custom-developed. (F p239)

Trends

100. Very-high-level languages will enable programmers to write programs with fewer statements than with today's high-level languages. (T p239)

101. High-level language machines will run high-level language programs directly without the need for compiling. (T p239)

102. Concurrent programming languages will allow programmers to specify that many tasks are to be performed simultaneously. (T p239)

103. Future computers are likely to be completely programmed in English. (F p239)

104. English is ideal language for programming computers. (F p239)

A Brief Introduction to LOGO

105. LOGO is a weak programming language, but it is nevertheless useful for children. (F p239)

106. The LISP programming language introduced the notions of turtle graphics. (F p239)

107. The following sequence of LOGO commands would produce a square:

```
FORWARD 50, RIGHT 90,
FORWARD 50, LEFT 90,
FORWARD 50, RIGHT 90,
FORWARD 50, LEFT 90.
```

(F p240)

108. The following LOGO procedure would draw a square:

```
TO DRAW
REPEAT 4 [FORWARD 50 RIGHT 90]
END
```

(T p240)

109. The following LOGO procedure could be used to draw a square 30 turtle steps on a side:

```
TO SQUARE: SIZE
REPEAT 4 [FORWARD: SIZE RIGHT 90]
END
```

(T p241)

Fill-in-the-Blanks

Programming Languages

Machine Language

1. Each computer can directly understand only one language--its own **machine language** .

2. Because machine languages are closely related to the structure of a particular computer, they are said to be **machine dependent** .

3. Programs written in machine language are not **portable** ; they may not be run on other computers with different machine languages.

Enrichment Piece: Telling a computer what to do

4. The statement, "I WANT A REPORT FOR BUSINESS TYPE 17 IN ORDER BY BRANCH, SHOWING ACCOUNT NUMBER, NAME, AND BOTH CURRENT AND PRIOR BALANCES." was probably written in **query** language.

Assembly Language

5. **Assembly** languages are "lower" than high-level languages, but "higher" than machine languages.

6. **Mnemonics** are English-like abbreviations used by programmers as memory aids.

7. **Assemblers** are computer programs or people that take programs written in mnemonics and translate them into machine language.

8. An assembler program is an example of a **translator** program.

9. The program of instructions written in assembly language is known as the **source** program.

10. An assember program translates a source program into a machine language program called an **object** program.

Macro Instructions

11. A **macro instruction** is one assembly language instruction that is translated into several machine language instructions.

12. Whenever the assembler program encounters a macro instruction, it first performs a **macro expansion** .

High-Level Languages

13. With **high-level** languages, programmers code programs that look like everyday English.

14. The translator programs that convert high-level language programs into machine language are called **compilers** .

15. An alternative to a compiler is an **interpreter** , a translator program that actually runs a source program directly.

Procedure-Oriented and Problem-Oriented Languages

16. **Procedure-oriented** languages are general purpose--they may be used to solve many different types of problems.

17. **Problem-oriented** languages are designed to solve specific types of problems.

"Quick-and-Dirty" Compilers and Optimizing Compilers

18. In **program development** environments, programmers are more concerned with debugging programs quickly than with having them run efficiently.

19. When a program is being debugged, **"quick-and-dirty"** compilers are used.

20. **"Quick-and-dirty"** compilers translate a program quickly, but the machine language programs they produce may not run as fast as possible and may use memory inefficiently.

21. When a program has been debugged it is put into **production** , that is, run on a regular schedule.

22. In **production** environments, programs must run efficiently.

23. In production environments, **optimizing** compilers are used.

A Survey of High-Level Languages

BASIC

24. The **BASIC** language was developed by John Kemeny and Thomas Kurtz.

25. The **BASIC** language was originally developed for interactive use on timesharing computers.

26. The **REM** statement in a BASIC language program indicates a remark.

FORTRAN

27. The oldest high-level language still in regular use is **FORTRAN** .

28. The **FORTRAN** programming language was developed by IBM in the mid-1950s.

29. The first programming language to undergo standardization was **FORTRAN.**

30. **The American National Standards Institute (ANSI)** is a government agency that sets standards in the United States.

COBOL

31. The **COBOL** programming language was developed for business data processing applications that must efficiently handle large volumes of data and edit it for presentation.

32. The **COBOL** programming language was developed by the Conference on Data Systems Languages.

33. Grace Hopper is associated with the development of the **COBOL** high-level programming language.

34. The part of a COBOL program that includes the author's name and the date the program was compiled is the **identification** division.

35. The **data** division of a COBOL program specifies the format of each of the data items to be used by the program.

36. The **procedure** division of a COBOL program specifies the actual calculations to be

performed.

37. COBOL tends to be **self-documenting** , requiring fewer additional comments and remarks than assembly language programs.

38. **Hierarachical data structuring** refers to the ability of a COBOL program to specify the structure of data in increasing levels of detail.

PL/1

39. The **PL/1** high-level programming language was developed by IBM for its System/360 series of computers.

40. **PL/1** is a structured high-level programming language that contains the better features of both FORTRAN and COBOL.

RPG

41. RPG stands for **Report Program Generator** .

42. With the RPG programming language, the programmer fills in **specifications** forms.

43. RPG is actually a program **generator** rather than a simple compiler that converts programs from high-level language to machine language.

APL

44. **APL** is a highly mathematical language requiring the use of many special symbols.

45. The **APL** programming language has a character set different from ASCII and requires a special keyboard.

Pascal

46. The programming language developed by Nicklaus Wirth in 1971 is **Pascal** .

47. **Pascal** is a descendant of ALGOL, a high-level language originally developed and used in Europe in the early 1960s.

48. Most Pascal compilers do not generate machine language instructions; rather they produce instructions in **p-code** .

C

49. The **C** programming language was used to code the UNIX operating system.

50. The programming language developed at Bell Laboratories and currently extremely popular among UNIX users is **C**.

Ada

51. The development of the **Ada** programming language was sponsored and coordinated by the U.S. Defense Department.

52. Pascal was used as the base for the **Ada** programming language.

53. The ability of Ada to allow a programmer to specify that many activities are to be performed simultaneously is called **multitasking** .

Forth

54. **Forth** , developed by Charles Moore, is not a particularly user-friendly language; it allows a programmer to gradually build up an ever richer set of capabilities.

LISP

55. The **LISP** programming language was developed by John Mc Carthy at the Massachusetts Institute of technology.

56. LISP stands for **LISt Processor** .

57. One of the first symbol manipulation languages was **LISP** .

58. One of the first languages to be used in text processing was **LISP** .

59. **LISP** is the preferred language in the field of artificial intelligence.

Query Languages

60. With **Query** languages computer users may request information from computers by typing simple English sentences.

A Universal Programming Language

61. If a **universal** programming language (UPL) were developed all programs would be portable.

62. If a **universal** programing language (UPL) were developed a huge amount of working software would have to be replaced.

Software Packages

63. **Packaged** software is used more than custom-developed software on personal computers.

64. With **packaged** software, users can obtain needed capabilities immediately without the problems and delays of developing their own software.

Trends

65. **Very-high-level** languages will enable programmers to write programs with fewer statements.

66. **High-level language machines** will run high-level language programs directly without the need for compiling.

67. **Concurrent programming** languages allow a programmer to specify that many tasks may to be performed simultaneously.

A Brief Introduction to LOGO

68. The programming language that is often associated with computers and children is **LOGO**.

69. The **LOGO** programming language was developed by Seymour Papert at MIT.

70. The graphics capabilities of LOGO are called **turtle** graphics.

Enrichment Piece: Building Turtle Designs with Procedures

71. The power of programming in LOGO comes when you teach the computer a new command, or more properly, a **procedure** .

72. The process, in the LOGO language, in which a procedure calls another version of itself is known as **recursion** .

Chapter 10
Systems Analysis and Design

Multiple Choice

Why Develop or Modify Systems?

1. An organization's set of systems is sometimes called its ______.
   (a) management information system *
   (b) pilot project
   (c) systems documentation
   (d) processing controls
   (e) computer auditing

The Systems Life Cycle

2. The series of phases a system goes through to make changes or modifications is called the ______.
   (a) systems performance audits
   (b) management information system
   (c) systems life cycle *
   (d) systems flowchart
   (e) systems analysis

The Requirements Definition Phase

3. When a system needs to be replaced or modified, users are asked to list their needs in the ______ phase.
   (a) systems analysis
   (b) systems design
   (c) operation
   (d) requirements definition *
   (e) implementation

4. The ______ is a list that provides management with information indicating what is needed, when it is needed, and why it is needed.
   (a) feasibility study
   (b) requirements document *
   (c) specification
   (d) cost/benefit analysis

(e) decision table

The Systems Analysis Phase

5. The requirements document is translated by systems analysts into a more precise document called the ______.
   (a) cost/benefit analysis
   (b) decision tree
   (c) feasibility study
   (d) systems documentation
   (e) specification *

6. Systems analysts act as intermediaries between users, management, and designers in the ______ phase.
   (a) implementation
   (b) requirements definition
   (c) systems analysis *
   (d) systems design
   (e) operation

The Feasibility Study

7. In the ______ the analyst examines the users' operations and how they relate to the rest of the organization.
   (a) feasibility study *
   (b) cost/benefit analysis
   (c) hierarchical module structure
   (d) pilot project
   (e) acceptance testing

8. ______ is how the user's operations relate to the rest of the organization.
   (a) Coupling
   (b) Interface *
   (c) Cohesion
   (d) Modules
   (e) Conversion

9. Which of the following is not stated in the specification? ______.
   (a) current problems
   (b) alternate solutions
   (c) costs and benefits
   (d) audit techniques *
   (e) implementation schedule

Data Collection

10. The collecting of ______ is the most costly, time-consuming, and error prone operation.
    (a) stubs
    (b) output controls
    (c) data flow
    (d) condition entries
    (e) input data *

Involving the Users

11. The systems analyst should especially involve the ______ throughout the systems development.
    (a) auditors
    (b) operators
    (c) users *
    (d) managers
    (e) owners

The Systems Design Phase

12. The ______ phase specifies how the system should be built.
    (a) systems design *
    (b) implementation
    (c) operation
    (d) systems analysis
    (e) requirements definition

Systems Flowcharts

13. ______ show how information flows through an organization.
    (a) Branches
    (b) Systems flowcharts *
    (c) Audits
    (d) Test cases
    (e) Conversions

Top-Down Stepwise Refinement

14. Structured design starts with the ______ process.
    (a) pseudocode

(b) hierarchical model structure
(c) data flow diagram
(d) top-down stepwise refinement *
(e) module

15. The breaking down of the top in top-down stepwise refinement creates many ______.
(a) lurking bugs
(b) output controls
(c) conversions
(d) audit trails
(e) modules *

16. The stepwise refinement process yeilds a ______.
(a) test team
(b) audit trail
(c) hierarchical module structure *
(d) phased conversion
(e) bottleneck

Cost/Benefit Analysis

17. In the ______ comparisons are made to what the organization should expect from the new system and what the prices are.
(a) implementation phase
(b) cost/benefit analysis *
(c) change control board
(d) remote diagnosis
(e) processing controls

Controls

18. Systems are designed with many ______ to enable them to check their operations.
(a) controls *
(b) audits
(c) modules
(d) test cases
(e) branches

19. ______ should be designed into the system, not as an afterthought.
(a) Branches
(b) Audits
(c) Test cases
(d) Controls *
(e) Modules

20. The highlighting of information systems personnel is called ______.
    (a) auditing around the computer
    (b) unit testing
    (c) management by exception *
    (d) maintenance
    (e) computer auditing

21. ______ checks determine if values fall within acceptable limits.
    (a) Reasonableness
    (b) Validity
    (c) Record
    (d) Batch
    (e) Range *

22. The checks that are used to determine if values are suspecious are ______ checks.
    (a) validity
    (b) reasonableness *
    (c) batch
    (d) range and limit
    (e) record

23. ______ checks determine if the inputs are in the proper format.
    (a) Record
    (b) Batch
    (c) Validity *
    (d) Range and limit
    (e) Reasonableness

24. ______ ensure that files are complete.
    (a) Record counts *
    (b) Reasonableness checks
    (c) Batch totals
    (d) Validity checks
    (e) Range and limit checks

25. To ensure that a group of transactions is complete, ______ are used.
    (a) range and limit checks
    (b) validity checks
    (c) record counts
    (d) batch totals *
    (e) reasonableness checks

26. The ______ examines batches for errors.
    (a) auditor
    (b) input control clerk *

(c) systems analyst
(d) systems designer
(e) facilities manager

The Implementation Phase

27. When the proposed design is accepted by management it is turned over to the ______, which acquires the needed hardware.
(a) systems designer
(b) facilities manager
(c) test team
(d) auditor
(e) implementation team *

28. In the ______ phase the system is tested and corrected until it works properly.
(a) requirements definition
(b) systems analysis
(c) implementation *
(d) systems design
(e) operation

29. When a system is run on a regular schedule it is said to be in ______.
(a) production *
(b) conversion
(c) acceptance testing
(d) auditing
(e) maintenance

Hardware Acquisition

30. A disadvantage to ______ is that when new equipment is wanted it may be difficult to dispose of the original equipment.
(a) leasing
(b) coupling
(c) purchasing *
(d) turnkey systems
(e) renting

31. Inflation encourages ______ because the fixed payments can be made with ever-cheaper dollars.
(a) turnkey systems
(b) leasing *
(c) facilities management
(d) purchasing

(e) renting

32. With ______ the customer pays a fixed monthly rate and builds up no ownership credits.
   (a) purchasing
   (b) turnkey systems
   (c) facilities management
   (d) renting *
   (e) leasing

Modules, Coupling, and Cohesion

33. Interacting groups of program pieces are called ______.
   (a) logic
   (b) branches
   (c) decision trees
   (d) decision tables
   (e) modules *

34. Modules should not be closely ______.
   (a) branched
   (b) cohesioned
   (c) coupled *
   (d) stubed
   (e) leafed

35. With strong ______, functions placed within a single module are as closely related as possible.
   (a) cohesion *
   (b) logic
   (c) coupling
   (d) documentation
   (e) branching

Training the Personnel

36. ______ can be a slow and costly process but it is essential for users who wish to understand the system.
   (a) Coupling
   (b) Audit trails
   (c) Conversion
   (d) Training *
   (e) Systems performance audits

Documentation

37. The ______ describes the state of the system at any given time in its development.
    (a) audit
    (b) documentation *
    (c) conversion
    (d) logic
    (e) data dictionery

38. The ______ documentation tells how to use the system.
    (a) operations
    (b) systems
    (c) user *
    (d) audit
    (e) management

39. The ______ documentation tells how to set up and run the programs.
    (a) user
    (b) management
    (c) operations *
    (d) audit
    (e) systems

40. The ______ documentation describes how the system is implemented.
    (a) systems *
    (b) audit
    (c) management
    (d) operations
    (e) user

Testing and Debugging

41. The process of demonstrating that the system operates according to its specification is called ______.
    (a) debugging
    (b) testing *
    (c) production
    (d) auditing
    (e) conversion

42. ______ are designed to check out the new system.
    (a) Validity checks
    (b) Reasonableness checks
    (c) Computer audits
    (d) Test cases *

(e) Audit trails

43. The ______ writes special software that checks that the system meets its specifications.
    (a) facilities manager
    (b) auditor
    (c) test team *
    (d) systems designer
    (e) consultant

44. Removing errors from a system is called ______.
    (a) auditing
    (b) testing
    (c) conversion
    (d) limit checking
    (e) debugging *

45. ______ is a term used for problems that can arise long after a system is installed.
    (a) Lurking bugs *
    (b) Condition entries
    (c) Stub delay
    (d) Bottleneck latency
    (e) Remote diagnosis

46. In ______ testing, each module is tested by itself.
    (a) acceptance
    (b) unit *
    (c) batch
    (d) integration
    (e) operation

47. With ______ testing a module is incorporated into the current complete version of the system.
    (a) operation
    (b) unit
    (c) acceptance
    (d) integration *
    (e) batch

48. Flaws in the manner in which modules communicate with one another are called ______.
    (a) process explosions
    (b) remote diagnosis
    (c) lurking bugs
    (d) condition entries
    (e) interface problems *

49. In ______ testing users test the system in a production environment.

(a) operation
(b) batch
(c) acceptance *
(d) unit
(e) integration

The Operation Phase

50. The ______ phase requires the careful attention of users, management, and the systems staff to ensure that the system is functioning properly.
(a) operation *
(b) implementation
(c) systems analysis
(d) systems design
(e) requirements definition

Converting to a New System

51. The changing over from the old system to the new one is called ______.
(a) branching
(b) conversion *
(c) implementation
(d) production
(e) acceptence testing

52. In ______ conversion both the new and the old system are run together for a period of time.
(a) crash
(b) phased
(c) parallel *
(d) pilot
(e) operation

53. In ______ conversion, the old system is dropped and only the new system is run.
(a) pilot
(b) parallel
(c) operation
(d) phased
(e) crash *

54. In ______ conversion the new system is installed a portion at a time.
(a) crash
(b) operation
(c) parallel
(d) phased *

(e) pilot

55. In ______ conversion only a small portion of the business is converted over to the new system at a time.
    (a) pilot *
    (b) phased
    (c) operation
    (d) parallel
    (e) crash

56. In a ______ generally the best branch of a company is converted to the new system first.
    (a) crash conversion
    (b) pilot project *
    (c) turnkey system
    (d) parallel conversion
    (e) phased conversion

Evaluation: The Systems Performance Audit

57. ______ audits determine if the system is operating properly.
    (a) Software
    (b) Service
    (c) Systems performance *
    (d) Operations
    (e) Conversion

58. ______ are hardware, software, or people who cannot handle the volume of work directed their way.
    (a) Remote diagnosis
    (b) Facilities management
    (c) Data flow diagram
    (d) Bottlenecks *
    (e) Procedure call

Maintenance

59. The ______ considers proposals and determines which ones should be considered.
    (a) change control board *
    (b) consultant
    (c) facilities manager
    (d) systems analyst
    (e) systems designer

Computer Auditing

60. _____ involves a person or another system tracing outputs back through the system to the inputs that caused them.
   (a) Auditing through the computer *
   (b) Data flow diagrams
   (c) Auditing around the computer
   (d) Remote diagnosis
   (e) Documentation

61. Any modifications must be _____, or recorded for later reference.
   (a) secured
   (b) logged *
   (c) documented
   (d) checked
   (e) totaled

62. With _____ an auditing program is generally used.
   (a) audit trails
   (b) remote diagnosis
   (c) auditing through the computer *
   (d) data flow diagrams
   (e) auditing around the computer

63. _____ checks the processing that occurs outside the computer.
   (a) Auditing through the computer
   (b) Audit trails
   (c) Auditing around the computer *
   (d) Data flow diagrams
   (e) Systems flowcharts

Service

64. Equipment that can check itself and in some cases make repairs is called _____.
   (a) immediate-response service
   (b) next-day service
   (c) remote diagnosis
   (d) self-checking hardware and software *
   (e) four-hour-response service

65. _____ refers to the ability of a company to check equipment over the telephone.
   (a) Four-hour-response service
   (b) Remote diagnosis *
   (c) Next-day service
   (d) Immediate-response service

(e) Self-checking hardware and software

66. ______ service is generally very expensive.
   (a) Immediate-response *
   (b) Next-day
   (c) Four-hour-response
   (d) Self-checking hardware and software
   (e) Remote diagnosis

67. A moderately priced service is ______ service.
   (a) remote diagnosis
   (b) self-checking hardware and software
   (c) four-hour-response *
   (d) next-day
   (e) immediate-response

68. An economical service option is ______ service.
   (a) immediate-response
   (b) four-hour-response
   (c) remote diagnosis
   (d) self-checking hardware and software
   (e) next-day *

Systems Acquisition

Software Packages

69. A program or system produced by an individual or company for sale to another individual or company is called a ______.
   (a) service bureau
   (b) software package *
   (c) turnkey system
   (d) facilities manager
   (e) consultant

70. With ______, programs are ready for immediate use and the vendors supply all documentation.
   (a) software packages *
   (b) systems acquisition
   (c) pseudocode
   (d) structured specification
   (e) consultants

Service Bureaus

71. Companies that develop a centralized installation with hardware and software and make it available for a fee are called ______.
   (a) turnkey systems
   (b) consultants
   (c) facilities manager
   (d) service bureaus *
   (e) software packages

72. ______ enable users to obtain computing capabilities without the hassles of hardware and software acquisition.
   (a) Software packages
   (b) Turnkey systems
   (c) Service bureaus *
   (d) Consutants
   (e) Facilities management companies

Turnkey Systems

73. A ______ is a complete computing system supplied by a single vendor but run by company people.
   (a) facilities manager
   (b) consultant
   (c) service bureau
   (d) software package
   (e) turnkey system *

Facilities Management

74. A ______ is hired by a client to run the company's entire data-processing operation.
   (a) facilities manager *
   (b) consultant
   (c) service bureau
   (d) software package
   (e) turnkey system

75. With ______ a company relinquishes control of its data-processing operations.
   (a) turnkey systems
   (b) facilities management *
   (c) consultants
   (d) software packages
   (e) service bureaus

Consultants

76. Experienced people who are hired to assist with various phases of the systems development and acquisition process are called ______.
    (a) auditors
    (b) systems analysts
    (c) consultants *
    (d) turnkey suppliers
    (e) systems designers

Case Study 10-1: Structured Systems Analysis and Design

77. A graphical model of how information moves through a system is called a ______.
    (a) data dictionary
    (b) process explosion
    (c) decision tree
    (d) data flow diagram *
    (e) decision table

78. The circle in a data flow diagram indicates ______.
    (a) the flow of information
    (b) a process
    (c) a data store
    (d) a decision table
    (e) a source or a sink *

79. The ______ contains a complete listing of all the components and data elements of the system.
    (a) data flow
    (b) data dictionary *
    (c) decision tree
    (d) process explosion
    (e) decision table

80. The ______ shows the detailed composition of all the data flows and data stores.
    (a) decision point
    (b) decision table
    (c) data dictionary *
    (d) decision tree
    (e) process explosion

81. In top-down stepwise refinement, the analyst performs a ______ to show more internal structure.
    (a) process explosion *

(b) decision point
(c) branch explosion
(d) data flow
(e) procedure call

82. The internal operation of a process is sometimes called ______.
(a) branching
(b) data flow
(c) leafing
(d) logic *
(e) procedure call

83. The decision chart that uses decision points and branches is a ______.
(a) process explosion
(b) decision tree *
(c) condition
(d) decision table
(e) procedure call

84. The decision chart that contains condition entries is a ______.
(a) procedure call
(b) process explosion
(c) decision tree
(d) subroutine
(e) decision table *

85. In a decision table the upper left portion lists the various ______.
(a) actions
(b) rules
(c) conditions *
(d) condition entries
(e) stubs

86. The bottom left portion of a decision table lists the ______.
(a) actions *
(b) stubs
(c) condition entries
(d) rules
(e) action entries

87. The vertical columns on the right of the decision table are called ______.
(a) drivers
(b) rules *
(c) actions
(d) conditions
(e) condition entries

88. With a hierarchical module structure, the boss acts in a management capacity and the workers do the real work of the system communicating via a ______.
    (a) subroutine
    (b) stub
    (c) driver
    (d) procedure call/return mechanism *
    (e) condition

89. A ______ is a hollow module that merely returns a preprogrammed result when called.
    (a) condition
    (b) driver
    (c) process
    (d) subroutine
    (e) stub *

90. A version of the boss module that calls worker modules with the specific information needed for test cases is called a ______.
    (a) stub
    (b) condition
    (c) driver *
    (d) procedure call/return mechanism
    (e) subroutine

True/False

1. The accounts payable system is designed to pay other companies for goods and services they supply. (T p246)

2. The accounts receivable system is designed to bill customers for goods and services they purchase. (F p246)

3. The order entry system is designed to keep track of all items in warehouses and on the shelves; when quantities get low it reorders. (F p246)

4. The sales analysis system prepares the income statements and balance sheets. (F p246)

Why Develop or Modify Systems?

5. An organization's set of systems is sometimes called its management information system. (T p247)

The Systems Life Cycle

6. The systems analysis phase begins after the systems design phase is completed. (F p247)

7. The operation and maintenance phase begins after the implementation phase. (T p247)

The Requirements Definition Phase

8. The users list and justify their new needs in the requirements definition phase. (T p247)

9. The requirements document is a general technical description of the system to be implemented. (F p247)

10. Commitments to change a system or develop a new one are not to be taken lightly. (T p247)

The Systems Analysis Phase

11. Systems analysts translate the specification into a requirements document. (F p248)

12. Systems analysts act as intermediaries between users, management, and systems designers. (T p248)

The Feasibility Study

13. The systems analyst proceeds with a feasibility study to determine if a system can be built to satisfy the users needs, and if it can be built within a reasonable amount of time and at acceptable costs. (T p248)

14. The relationship between the user's operations and the rest of the organization is an example of an interface. (T p248)

15. Written questionnaires are routinely used by systems designers as they survey user needs. (F p248)

16. An important part of the systems analyst's task is to evaluate many alternative approaches. (T p248)

17. It is important that the systems analyst discuss all alternative approaches that were considered, and the analyst should document these carefully in the specification. (F p248)

18. The specification produced by the systems analyst should be a highly detailed document that convincingly demonstrates both the feasibility of the system and how to design it. (F p248)

19. The level of detail that should be used in the specification depends a great deal on the relationship between the systems analysts and the systems designers. (T p248)

20. The interface between systems analysts and systems designers is not crisply defined and often varies among organizations. (T p249)

Data Collection

21. Collecting input data is one of the more economical problems in computer-based systems. (F p249)

22. Often a company that has computerized some of its operations will discover that it is already collecting and computerizing most, and possibly all, of the data it needs to feed a new system. (T p249)

Involving the Users

23. In general, the users should be carefully consulted with during the systems analysis phase, so that there is no need to involve them throughout the remainder of the systems development effort. (F p249)

The Systems Design Phase

24. The specification that emerges from the systems analysis phase describes *what* the desired system is to do. In the systems design phase, the systems designers decide *how* to do it. (T p249)

Systems Flowcharts

25. A systems flowchart indicates how information flows between the various components of a system. (T p249)

26. Structured program flowcharts use a much larger variety of symbol shapes than systems flowcharts do. (F p249)

Top-Down Stepwise Refinement

27. Structured design begins with the top-down stepwise refinement process. (T p249)

28. The top-down stepwise refinement process terminates when the number of modules becomes

too large. (F p249)

29. The stepwise refinement process yields a hierarchical module structure. (T p249)

30. In a typical hierarchical structure each employee has only one boss, but each boss may have many employees. (T p249)

31. In a hierarchical module structure, each module performs its function only when called on by the module immediately below it in the structure. (F p249)

32. The systems flowchart symbol which appears to be a rectangle with its top-left corner sliced off at an angle designates manual input. (F p250)

Cost/Benefit Analysis

33. Benefits such as "making information more accessible" are more difficult to measure than benefits such as "50 percent faster processing." (T p251)

34. In general, controls should be added to a working system only when the need for them becomes apparent. (F p252)

35. Processing controls prevent invalid data from entering the system. (F p252)

36. Highlighting suspicious information for systems personnel to investigate is called management by suspicion. (F p252)

37. Checking that a date in November must be between the 1st and the 30th is an example of a reasonableness check. (F p252)

38. Range and limit checks are used to determine if certain values are suspicious. (F p252)

39. A reasonableness check would call attention to a person's age of 130. (T p252)

40. When a data item fails a range check, it must be incorrect. (T p252)

41. When a data item fails a reasonableness test, it must be incorrect. (F p252)

42. Validity checks determine if the input is in the proper format. (T p252)

43. A limit check would determine that a numeric field containing a question mark is incorrect. (F p252)

44. Record counts are used as controls to help ensure that files are complete. (T p252)

45. Batch totals help ensure that a group of transactions is complete. (T p252)

The Implementation Phase

46. The needed hardware for a new system is acquired during the design phase. (F p252)

47. A new system is programmed during the implementation phase. (T p252)

48. Once it has been demonstrated that a new system is functioning properly, the system enters production. (T p252)

49. It is important for the systems designers to specify the design in a manner such that the implementation team will have great flexibility in building the system. (F p252)

Hardware Acquisition

50. The three popular ways of acquiring hardware are purchasing, leasing, and renting. (T p253)

51. Inflation discourages the leasing of computer hardware. (F p253)

52. As a rule, if the computer will not be kept for a long period such as five or more years, it is better to lease than to buy. (T p253)

53. Many leases contain purchase options; the longer the computer hardware has been leased when the decision is made to purchase it, the lower the purchase price. (T p253)

54. When equipment is rented, the risks are higher than with leasing, but the costs are lower. (F p253)

55. Rental agreements are usually for a relatively short term. (T p253)

56. Normally, with rental agreements the user builds up no ownership credits. (T p253)

Modules, Coupling, and Cohesion

57. Most of today's systems are built as groups of interacting program pieces called modules. (T p253)

58. The "divide and conquer strategy" is an argument against breaking a system into modules. (F p253)

59. The more detail in a design, the more difficult it is for programmers to translate the design into working programs. (F p253)

60. Deciding how to partition a system into modules is a relatively easy task but nevertheless an important one. (F p253)

61. Coupling is a measure of the interdependence of modules. (T p253)

62. Coupling should be minimized. (T p253)

63. Cohesion is a measure of the relationship between functions placed in a single module. (T p253)

64. Cohesion should be minimized. (F p253)

65. If modules are not closely coupled, then unanticipated side effects often result. (F p253)

66. If modules are closely coupled, it is likely that when one bug is fixed new ones may appear. (T p253)

Training the Personnel

67. Training, although time-consuming and costly, is essential; the users will accept and benefit by a system more readily if they understand it. (T p254)

Documentation

68. Systems documentation tells the computer operators how to set up and run the programs. (F p254)

69. Operations documentation is important for modifying the system. (F p254)

70. User documentation describes how the system is actually implemented. (F p254)

Testing and Debugging

71. Testing is the process of demonstrating that the system operates according to its specification. (T p254)

72. For efficiency, the test team should consist of people who wrote the software. (F p254)

73. The test team must check every possibility to be sure that the system is functioning properly. (F p254)

74. Locating and removing bugs is a simple but time-consuming task. (F p254)

75. Removing a bug exposed by a test case guarantees that the program will now pass that test case. (F p255)

76. Lurking bugs will sometimes surface years after a system has been installed. (T p255)

77. Modules are unit tested before they undergo integration testing. (T p255)

78. In unit testing, a module is integrated into the current working version of the system to see if it indeed functions properly. (F p255)

79. One goal of unit testing is to expose interfacing problems between modules. (F p255)

80. Systems contracts generally provide for a period of acceptance testing during which users test the system in a production environment to determine if it does indeed fulfill their requirements. (T p255)

Converting to a New System

81. Changing from an old system to a new system is called conversion. (T p255)

82. In a parallel conversion, only a small portion of the business is converted over to the new system at once. (F p255)

83. Parallel conversion is a conservative approach. (T p255)

84. Parallel conversions are costly because they require people and resources to run two systems. (T p255)

85. The most dramatic conversion technique is the phased conversion. (F p255)

86. A crash conversion can result in the most economical conversion of all. (T p255)

87. A crash conversion can be devastating. (T p255)

88. The crash conversion is a popular technique because of its speed and economy. (F p255)

89. In a phased conversion, the new system is phased in a portion at a time. (T p255)

90. A phased conversion requires fewer resources than a parallel conversion. (T p255)

91. In a pilot conversion, the system is phased in a small portion at a time on the pilot project. (F p255)

92. When a company with 10 branch offices converts only the best run branch as an experiment

this is called a pilot project. (T p255)

Evaluation: The Systems Performance Audit

93. Problems normally arise even when carefully designed and implemented systems are put into production. (T p256)

94. Bottlenecks are hardware, software, or people who cannot handle the volume of work directed their way. (T p256)

95. Removing a bottleneck immediately improves performance. (F p256)

Maintenance

96. The job of a change control board is to test and debug all changes to a system. (F p256)

97. Many organizations allocate more than half of their software budgets to maintaining existing systems. (T p256)

Computer Auditing

98. An audit trail allows a person to trace outputs back through the system to the inputs that caused them. (T p256)

99. The auditor must reconstruct all computer processing to prove that the system is operating properly. (F p256)

100. Auditing through the computer requires examining the detailed processing of data inside the computer system. (T p256)

101. Auditors who can audit through the computer are common because of the large number of courses offered in computing and data processing. (F p256)

102. Auditing around the computer means that the auditor traces only the processing that occurs outside the computer. (T p256)

103. Auditing through the computer requires more skill than auditing around the computer. (T p256)

104. Auditing around the computer is the key to detecting and blocking today's sophisticated computer criminals. (F p256)

Service

105. Most computers today repair themselves. (F p256)

106. Preventative maintenance helps reduce the risk of hardware breakdowns. (T p256)

107. Service is normally included with leased equipment, but not with rented equipment. (F p256)

108. Service on systems configured from the equipment of several vendors is not much different from service on systems configured from a single vendor's equipment. (F p257)

109. Transferring a disgnostic program from the service company's computer to the client's computer over the telephone is called uploading. (F p257)

Systems Acquisition

110. The vast majority of businesses today develop all of their own systems to meet their own needs. (F p257)

Software Packages

111. In the past, most of a company's software systems were "home-grown." (T p257)

112. The number of high-quality software packages on the market today is still very limited and these tend to be priced rather high. (F p257)

113. Software packages generally cost 10 times as much as home-grown software. (F p257)

114. One advantage of using software packages over home-grown software is that they are bug free. (F p257)

115. A disadvantage of software packages is that they cannot be modified. (F p257)

116. Service contracts are available for software as well as hardware. (T p257)

Service Bureaus

117. Service bureaus enable users to obtain computing capabilities without the hassles of hardware and software acquisition. (T p258)

118. A company with its own computers has no need to use a service bureau. (F p258)

Turnkey Systems

119. With a turnkey system, the vendor normally runs the system with its own people. (F p258)

120. One problem with a turnkey system is the long period of debugging that generally occurs after the system is delivered. (F p258)

121. One benefit of using turnkey systems is that the gambles associated with developing home-grown systems are eliminated. (T p258)

122. A company with its own systems development group will not use turnkey systems. (F p258)

123. One problem with turnkey systems is that users may have to bend their procedures to fit the systems rather than the other way around. (T p258)

Facilities Management

124. A facilities management company is hired by a client to do it all--to create and run the client company's entire data processing operation. (T p258)

125. The main disadvantage of using facilities management is that the company is relinquishing control of its information lifeline. (T p258)

126. A company that already has a data processing operation will not hire a facilities manager. (F p259)

Consultants

127. Consultants are experienced people from outside the organization who are hired to make the decisions about the organization's data processing operations. (F p259)

Case Study: Structured Systems Analysis and Design

128. The analyst's job in structured systems analysis is to produce the structured systems specification. (T p260)

129. A data flow diagram is a graphical model of how data flows through a system. (T p260)

130. A data flow diagram clearly and precisely shows the decision-making operations in a system. (F p260)

131. The four symbols commonly used in data flow diagrams are for data flows, processes, data stores, and sources or sinks. (T p260)

132. In data flow diagrams, the same symbol is used for a source and a sink. (T p260)

133. A data flow is an individual piece of data that flows between components of a system. (F p260)

134. The system's analyst must carefully specify the exact form in which data is to be typed. (F p260)

135. The system's designer must carefully specify the exact form in which data is to be typed. (T p260)

136. In a data flow diagram, the open-sided rectangle indicates a process. (F p260)

137. In a data flow diagram, the shape of every symbol uniquely determines what type of system component it represents. (F p260)

138. In a data flow diagram, a circular symbol that has an arrow flowing into it must be only a sink. (F p261)

139. In a data flow diagram, a rectangular symbol that has an arrow flowing into it must be a process. (T p261)

140. In a data flow diagram, the only type of component that is considered to be external to the system is a source. (F p261)

141. In a data flow diagram, if a data store has an error entering it and an arrow leaving it, then something must be wrong. (F p261)

142. The systems analyst must clearly and precisely specify the format of the information in each data store. (F p261)

The Data Dictionary

143. The data dictionary is a complete listing of all the components and data elements of a system as defined in the data flow diagram. (T p262)

144. A data store must contain at least the information that is to be provided in the outward data flows. (T p262)

145. If a data store has no inward data flow, this may mean that the data store is replenished by some subsystem other than the one currently being specified. (T p263)

The Process Explosion

146. The "top" process in a data flow diagram is refined by "exploding" the process. (T p263)

147. A process is exploded to show more of its external structure. (F p263)

148. Each process may reference only one data store. (F p263)

Specifying the Operation of Processes

149. The internal operation of a process is sometimes called its "logic." (T p263)

150. A simple English narrative is not precise enough to specify the internal operation of a process. (F p263)

A Decision Tree

151. The branches in a decision table are followed to arrive at key decision points. (F p263)

152. The end of each path in a decision tree is called a stump. (F p263)

A Decision Table

153. The upper left portion of the decision table lists the various conditions to be considered. (T p264)

154. The bottom left portion of the decision table lists the actions that are to be taken. (T p265)

155. The right half of the decision table is divided into a series of vertical columns called rules. (T p265)

156. The top portion of each rule contains the action entries. (F p265)

157. The condition entries must be "Yes" or "No" indicating which conditions apply to a particular rule. (F p265)

158. A dash in a condition entry means this condition should be ignored. (T p266)

159. The bottom half of each rule contains the condition entries. (F p266)

160. An "X" in an action entry indicates that the action is not to be taken. (F p266)

161. Dashes may be used in condition entries and in action entries. (T p266)

Pseudocode

162. With pseudocode, it is possible to specify the same logic as with decision tables. (T p266)

163. With pseudocode, it is not possible to specify the same logic as with decision trees. (F p266)

Beginning the Structured Design

164. The structured design process is concerned with translating a physical model of the system into a logical model of the system. (F p266)

165. A logical model of a system shows what the system does. (T p266)

166. A physical model of a system describes how the system is to be built and how it is to operate. (T p266)

Hierarchical Module Structure

167. A system should be modularized in such a manner as to facilitate future modifications. (T p266)

168. The choice of how to modularize a system should be made independently of the data flow diagram. (F p266)

169. The real work in a modularized system is performed by the "boss" modules. (F p266)

170. In a hierarchical module structure, each "boss" module is a relatively sparse shell that calls the worker modules in appropriate sequence. (T p266)

171. "Boss" modules must call the "worker" modules sequentially. (F p266)

Stubs and Drivers

172. Once a "boss" module is implemented, it can be tested by writing stubs for each of its "worker" modules. (T p266)

173. A stub is a "hollow" module that merely returns a preprogrammed result when called. (T p266)

174. If a few of the "worker" modules have been implemented, they may be tested by writing a version of the "boss" module called a driver module that simply calls the "worker" modules with the specific information needed for these test cases. (T p268)

175. A stub module is used to test "worker" modules. (F p268)

176. Driver modules are used to test a "boss" module. (F p268)

Fill-in-the-Blanks

1. A computer-based system is one with "things or parts" that consist primarily of **hardware** and software that work together to perform some common goal.

2. Seeing how various components may be combined to form systems is called the **systems** view.

3. The system that pays other companies for goods and services they supply is the **accounts payable** system.

4. The **accounts receivable** system keeps track of money owed by customers and sends regular statements requesting payment.

5. The **order entry** system accepts customer orders for goods and services.

6. The system that keeps track of all items on the shelves and in warehouses is the **inventory control** system.

7. The system that keeps track of a company's profit and loss as well as its assets and liabilities is the **general ledger** system.

Why Develop or Modify Systems?

8. An organization's set of systems is sometimes called its **management information system (MIS)**.

The Systems Life Cycle

9. As new systems are developed or existing ones are changed, they evolve through a series of phases called the **systems life cycle**.

The Requirements Definition Phase

10. The listing of reasons justifying a new system are expressed in the **requirements definition** phase.

11. The list produced during the requirements definition phase is called the **requirements document** .

The Systems Analysis Phase

12. The user's needs listed in the requirements are translated by **systems analysts** into a more precise document.

13. The **specification** describes the type of system that the analyst feels would be appropriate to suit the company's needs.

14. **Systems designers** use the specification as a basis for actually designing the system.

15. In the **systems analysis** phase, systems analysts act as intermediaries between users, management, and system designers.

The Feasibility Study

16. In the **feasibility** study, the analyst thoroughly examines the user's operations and how they relate to the rest of the organization.

17. It is important to study how the operations of an organization relate to, or **interface** with the rest of the organization.

Data Collection

18. The collecting of **input** data is one of the most costly, time-consuming, and error-prone operations in computer-based systems.

Involving the Users

19. It is important for the systems analysts to involve the **users** throughout the systems development effort.

The Systems Design Phase

20. The specification that emerges from the systems analysis phase describes **what** the desired system is to do.

21. In the **systems design** phase, the systems designers decide how to build the new system to do what the users need.

Systems Flowcharts

22. A **systems flowchart** indicates how information flows through the various components of the system.

Top-Down Stepwise Refinement

23. With the **top-down stepwise refinement** process, the design is developed gradually through a series of increasingly detailed refinements.

24. The first step in a top-down stepwise refinement is called the **top**.

25. The first refinement in the top-down stepwise refinement is to break the top into separate **modules**.

26. The stepwise refinement process yields a **hierarchical** module structure.

Cost/Benefit Analysis

27. The evaluation process that compares several systems by considering their costs and benefits to the company is called **cost/benefit analysis**.

Controls

28. Systems are designed with many **controls** to enable them to check their operations.

29. **Controls** should be designed into a system, not added as an afterthought.

30. When information is highlighted for investigation by personnel, this is called **management by exception**.

31. **Reasonableness** checks are used to determine if certain values are suspicious.

32. **Validity** checks determine if the input is in the proper format.

33. **Record** counts are used to ensure that files are complete.

34. **Batch totals** help ensure that a group of transactions is complete.

35. An **input control** clerk examines batches for errors.

The Implementation Phase

36. When the proposed design is accepted by management, it is turned over to the **implementation team**.

37. In the **implementation** phase, hardware is acquired, the system is programmed, tests are run to uncover bugs, and so on.

38. When a system is run on a regular schedule to produce useful work it is said to be in **production**.

Hardware Acquisition

39. A disadvantage to **purchasing** equipment is that it is often hard to dispose of the original equipment in the used computer marketplace.

40. With **leased** equipment it is possible for equipment to be upgraded.

41. When equipment is **rented** there is a fixed monthly cost but no ownership credits are built up.

Modules, Coupling, and Cohesion

42. Most systems are built as groups of interacting program pieces called **modules**.

43. **Coupling** should be minimized--modules should be as independent of one another as possible.

44. **Cohesion** should be maximized--functions placed within a single module should be as closely related to one another as possible.

Training the Personnel

45. Teaching the users how to operate the new system is done in the **training** process.

Documentation

46. **Documentation** provides the "user's view" of the system.

47. When a new system is under development, its **documentation** describes the state of the system at any given moment.

48. **User** documentation tells how to use the system.

49. **Operations** documentation tells the computer operators how to set up and run the programs.

50. **Systems** documentation describes how the system is actually implemented.

Testing and Debugging

51. **Testing** is the process of demonstrating that the system operates according to its specification.

52. The **test** team designs the test cases to check out the system.

53. In **unit** testing, each module is tested by itself to see if it runs properly.

54. In **integration** testing, a module is incorporated into the current working version of the system.

55. **Interfacing** problems are flaws in the manner in which modules communicate with one another.

56. During **acceptance** testing users test the system in production environments.

The Operation Phase

57. In the **operation** phase, the system goes into production.

Converting to a New System

58. Changing from the old system to the new one is called **conversion**

59. **Parallel** conversion is when both the old and new systems run together for some period of time.

60. In a **crash** conversion, the new system is used immediately and the old system is no longer used.

61. With a **phased** conversion, a new system is put into operation a portion at a time.

62. In a **pilot** conversion, only a small portion of a business is converted over to the new system.

Evaluation: The Systems Performance Audit

63. **Systems performance** audits are performed regularly to evaluate the system and determine if it is operating properly.

64. **Bottlenecks** are hardware, software, or people who cannot handle the volume of work directed their way.

Maintenance

65. Some organizations have a **change control** board, to consider all proposed changes and determine which ones should actually be implemented.

Computer Auditing

66. An **audit trail** allows a person or another system to trace outputs back through the system to the inputs that caused them.

67. With **auditing through the computer** , outputs are traced back to inputs through the processing inside the computer.

68. An auditor traces only the processing that occurs outside the computer when **auditing around the computer** .

Service

69. Vendors now build their machines with **self-checking** hardware and software that automatically diagnoses most problems, and in some cases, even repairs them.

70. Many service companies can perform **remote disgnostics** that check the status of a computer over the telephone.

71. **Immediate** -response service is the most expensive.

72. **Four-hour** -response service is moderately priced.

73. An economocal service is **next**-day service.

Systems Acquisition

Software Packages

74. A **software package** is a program or system produced by an individual or company for sale to other individuals or companies.

75. **Software packages** are more likely than most home-grown software to be bug-free.

76. Some vendors offer **software** service contracts, whereby they agree to provide regular program updating.

Service Bureaus

77. **Service bureaus** are companies that develop a centralized installation with hardware, software, and databases, and makes these available to users for a fee.

78. **Timesharing** service bureaus install terminals on the client's premises.

79. **Service bureaus** enable users to obtain computing capabilities without the hassles of hardware and software acquisition.

80. With **service** bureaus, first-time users can gain the benefits of computing quickly and economically.

81. Often **service** bureaus develop considerable expertise in specialized areas.

Turnkey Systems

82. A **turnkey** system is a complete computing system including hardware, software, documentation, and training supplied by a single vendor.

83. **Turnkey** systems provide users with ready-to-go complete hardware/software packages; the users operate the systems.

Facilities Management

84. A client can hire a **facilities management** company to create and run the company's entire data processing operation.

85. **Facilities** management is attractive to companies that need to have their data processing operations cleaned up.

Consultants

86. **Consultants** provide expertise and advice, but the client company still makes its own decisions.

87. **Consultants** are experienced people from outside the organization who are hired to assist with various phases of systems development.

Case Study: Structured Systems Analysis and Design

Requirements

88. A company states its needs in the **requirements** phase.

Systems Analysis

The Data Flow Diagram:

89. The analyst's job in structured systems analysis is to produce the **structured systems specification** .

90. The **structured systems specification** contains a clear and precise description of what the system is supposed to do.

91. The **data flow diagram** is a graphical model showing how information travels through the system.

92. The circle symbol in the data flow diagram is a **source** or sink.

93. In the data flow diagram, a rectangle indicates that certain **processing** is to be performed.

94. The three-sided rectangle in the data flow diagram is called a **data store** .

The Data Dictionary:

95. A complete listing of all the components and data elements of the system as defined in the data flow diagram is found in the **data dictionary** .

96. The **data dictionary** shows the detailed composition of all the data flows and data stores.

The Process Explosion:

97. To show more of the internal structure of a process, the analyst **explodes** the process to reveal its subprocesses.

Specifying the Operation Processes:

98. The internal operation of a process is sometimes called the **logic** of the process.

A Decision Tree:

99. A decision **tree** is interpreted left to right by following branches.

100. The decision tree ends at **leaves** .

A Decision Table:

101. The upper left portion of a **decision table** lists the various conditions to be considered.

102. The bottom left portion of a decision table lists the **actions** that are to be taken.

103. The right half of a decision table is divided into a series of vertical columns called **rules** .

104. The **condition** entries on a decision table are simply "Yes," "No," or "-."

105. The **action** entries on a decision table are indicated by an "X" or a "-".

Pseudocode:

Beginning the Structured Design:

Hierarchical Module Structure:

106. A boss module calls upon worker modules using a **procedure call/ return** mechanism.

Stubs and Drivers:

107. A **stub** is a hollow module that merely returns a preprogrammed result when called.

108. A **stub** tests a boss module.

109. Worker modules are tested by writing a **driver** module.

Chapter 11
Database Management Systems, Management Information Systems, Decision Support Systems

Multiple Choice

Data as a Business Resource

1. The collection of all the information a business maintains is sometimes called a ______
   (a) record
   (b) database *
   (c) file
   (d) field
   (e) byte

2. The process of keeping information up-to-date and correct is called ______
   (a) database integrity
   (b) redundancy
   (c) file maintenance *
   (d) dunning
   (e) goal seeking

3. When the same information appears in more than one file we have what is called ______
   (a) redundancy *
   (b) database integrity
   (c) goal seeking
   (d) dunning
   (e) file maintenance

4. ______ can be a useful security measure if data is lost.
   (a) Dunning
   (b) Goal seeking
   (c) File maintenance
   (d) Database integrity
   (e) Redundancy *

Database Management Systems

5. A ______ creates a database, keeps it up-to-date, and provides access to authorized users.

(a) MIS
(b) DSS
(c) DAS
(d) DBMS *
(e) EDP

6. When what is supposed to be in the database is there and what is not supposed to be there isn't we have ______
(a) database integrity *
(b) dunning
(c) redundancy
(d) file maintenance
(e) data independence

7. A ______ determines who may access the database, modify it, and add new relationships.
(a) network database
(b) database administrator *
(c) relational database
(d) management hierarchy
(e) hierarchical database

Hierarchical Database

8. In a ______ database records are arranged in a parent/child relationship.
(a) network
(b) management
(c) hierarchical *
(d) tactical
(e) relational

9. In a ______ database, searching is most rapid when searched from the top down.
(a) tactical
(b) management
(c) relational
(d) network
(e) hierarchical *

10. If a database is organized with ______, a single copy of data can be made available to answer many different types of queries.
(a) pointers
(b) tuples
(c) domains
(d) multiple views *
(e) data dictionary

Network Database

11. With a ______ database, a variety of questions can be answered without having to duplicate data.
   (a) network *
   (b) management
   (c) redundant
   (d) hierarchical
   (e) tactical

12. ______ are special data items inside the computer that correspond to the arrows in a diagram.
   (a) Domains
   (b) Pointers *
   (c) Flat files
   (d) Key fields
   (e) Relations

Relational Database

13. A ______ database is considered to be more user-friendly and stores information in tables.
   (a) management
   (b) hierarchical
   (c) relational *
   (d) tactical
   (e) network

14. In a relational database, tables are sometimes called ______.
   (a) domains
   (b) key fields
   (c) open items
   (d) flat files *
   (e) assets

15. Each row in a relation is called a record or a ______.
   (a) key field
   (b) flat file
   (c) domain
   (d) open item
   (e) tuple *

16. Fields in a relational database are also called ______.
   (a) flat files
   (b) domains *

(c) tuples
(d) key fields
(e) assets

17. Two relations may be combined or ______.
(a) modeled
(b) merged
(c) joined *
(d) projected
(e) selected

18. When some columns are extracted from a relation and used to form a new relation, the relation is said to be ______.
(a) projected *
(b) modeled
(c) joined
(d) selected
(e) dunned

19. In a relational DBMS, records may be ______ according to various user-specified criteria.
(a) joined
(b) dunned
(c) projected
(d) modeled
(e) selected *

Case Study 11-1: A Simple Relational DBMS

20. In a relational DBMS, a ______ is a domain used for associating items in one relation with items in another.
(a) data dictionary
(b) flat field
(c) tuple
(d) key field *
(e) pointer

Data Dictionary

21. Detailed descriptions of each item in the database are kept in the ______.
(a) key field
(b) data dictionary *
(c) flat file inventory
(d) tuple
(e) flat file

22. The name, type, and size of each item in a database can be found in the ______
    (a) flat file
    (b) DBMS inventory
    (c) data dictionary *
    (d) tuple
    (e) key field

Data Independence

23. The ability to modify the structure of the database without affecting existing programs that reference the database is called ______.
    (a) redundancy
    (b) structured modification
    (c) tactical planning
    (d) database integity
    (e) data independence *

Management Information Systems

24. Which of the following is not one of the four key functions performed by to managers?

    ______
    (a) planning
    (b) dunning *
    (c) controlling
    (d) organizing
    (e) directing

25. An early buzzword in the field of computerized business systems was ______.
    (a) EDP *
    (b) MIS
    (c) DSS
    (d) DBMS
    (e) DAS

26. The primary goal of ______ systems is to reduce clerical costs by computerizing the flow of business transaction data.
    (a) DSS
    (b) ATB
    (c) EDP *
    (d) MIS
    (e) DBMS

27. In management hierarchy, medium level managers deal with short term ______ planning.

(a) strategic
(b) day-to-day
(c) inventory
(d) tactical *
(e) model

28. Upper-level managers tend to deal with long-term ______ planning.
(a) day-to-day
(b) model
(c) tactical
(d) inventory
(e) strategic *

Case Study 11-2: An MIS for FutureStore

29. The ______ system accepts requests for mechandise and sets into motion the mechanisms to ensure delivery.
(a) inventory
(b) order entry *
(c) accounts receivable
(d) decision automation
(e) general ledger

30. The ______ system instructs warehousing to ship goods to clients.
(a) general ledger
(b) inventory
(c) order entry *
(d) accounts receivable
(e) sales analysis

31. With a(n) ______ system, the status of each customer order is monitored so inquiries can be answered.
(a) order entry *
(b) inventory
(c) accounts receivable
(d) general ledger
(e) billing

32. The amount of each item a company manufactures that is kept on hand is shown on a(n) ______.
(a) suspense listing
(b) payroll listing
(c) chart of accounts
(d) inventory listing *
(e) assets listing

33. The ______ system attempts to minimizes the amount of a product a company keeps on hand while at the same time attempting to minimize the chance of stockouts.
(a) order entry
(b) accounts receivable
(c) general ledger
(d) sales analysis
(e) inventory control *

34. In an inventory control system, when more inventory is needed the computer prints a ______
(a) suspense listing
(b) purchase order *
(c) bill of materials
(d) check register
(e) income statement

35. The ______ lists every item in inventory, the quantity sold for a period of time, and the dollar amount for each item.
(a) purchase order
(b) cash requirements report
(c) distribution-by-value report *
(d) vendor master report
(e) income statement

36. A ______ system interfaces with order entry to determine what products need to be produced and when.
(a) manufacturing resources planning *
(b) inventory control
(c) accounts payable
(d) sales analysis
(e) credit and collection

37. A ______ is a complete listing of all the parts needed to produce a specific item.
(a) capacity requirements report
(b) bill of materials *
(c) suspense listing
(d) inventory report
(e) distribution-by-value report

38. The ______ requirements report states what parts will be needed when.
(a) capacity
(b) asset
(c) inventory
(d) material *
(e) manufacturing

39. What production capabilities will be needed when are stated in the ______ requirement report.
(a) manufacturing
(b) asset
(c) material
(d) inventory
(e) capacity *

40. When an order has been shipped, the ______ system sends out an invoice listing what was sent.
(a) accounts payable
(b) payroll
(c) billing *
(d) accounts receivable
(e) sales analysis

41. The job of the ______ system is to handle most money collections routinely.
(a) payroll
(b) accounts receivable *
(c) accounts payable
(d) billing
(e) general ledger

42. In a(n) ______ accounts receivable system, incoming cash is applied either to a specific invoice or the oldest invoice.
(a) open item *
(b) aged trial balance
(c) balance forward
(d) invoicing
(e) profit and loss

43. With a(n) ______ accounts receivable system, payments received are credited to an account and any new purchases are added to the total still owed.
(a) profit and loss
(b) invoicing
(c) balance forward *
(d) aged trial balance
(e) open item

44. In the ______ report, the status of every customer is shown including payments, purchases and new totals.
(a) suspense listing
(b) cash requirements
(c) vendor master
(d) aged trial balance *

(e) profit and loss

45. Transactions that must be handled manually are shown on the ______.
(a) aged trial balance
(b) balance sheet
(c) profit and loss statement
(d) income statement
(e) suspense listing *

46. The ______ system maintains in-house records on each customer and decides if a customer should be allowed to make additional purchases.
(a) accounts receivable
(b) credit and collection *
(c) inventory control
(d) accounts payable
(e) payroll

47. The ______ system ensures that employees are paid on time.
(a) accounts payable
(b) credit and collection
(c) payroll *
(d) general ledger
(e) check register

48. The ______ is a listing of all checks paid by a company.
(a) printed check report
(b) chart of accounts
(c) cash requirements report
(d) check register *
(e) balance sheet

49. With ______, employees can have their pay placed into their bank accounts electronically by computer.
(a) direct deposit *
(b) check register
(c) payroll system
(d) cash requirements report
(e) general ledger system

50. The ______ system pays suppliers and takes advantage of discounts they might offer.
(a) accounts receivable
(b) payroll
(c) accounts payable *
(d) direct deposit
(e) cash requirements

51. The amount of money a company needs on a week-by-week basis is shown in the ______ report.
(a) profit and loss
(b) cash requirements *
(c) general ledger
(d) cash receipts journal
(e) income

52. The complete listing of all suppliers with their addresses and payment terms is located in the ______ report.
(a) vendor master *
(b) chart of accounts
(c) company
(d) accounts receivable
(e) accounts payable

53. The ______ system keeps track of a company's financial records.
(a) accounts payable
(b) payroll
(c) accounts receivable
(d) general ledger *
(e) aged trial balance

54. The listing of all the categories to which general ledger applies charges and payments is the ______.
(a) suspense listing
(b) accounts payable report
(c) profit and loss report
(d) balance sheet
(e) chart of accounts *

55. The ______ statement summarizes by category all the income and expenses the business had for a specific amount of time.
(a) assets
(b) liabilities
(c) profit and loss *
(d) balance
(e) ledger

56. Everything a business owns is called its ______.
(a) liabilities
(b) income
(c) inventory
(d) assets *
(e) credit

57. ______ are what a business owes.
   (a) Assets
   (b) Credits
   (c) Liabilities *
   (d) Sales
   (e) Inventory items

58. A company's net worth is summarized in the ______.
   (a) balance sheet *
   (b) inventory listing
   (c) distribution-by-value
   (d) payroll
   (e) chart of accounts

59. An auditor will give a(n) ______ opinion if a company's books are in good order.
   (a) adverse
   (b) unqualified *
   (c) mathematical
   (d) ad hoc
   (e) qualified

60. If certain areas of a financial report are weak, an auditor will issue a(n) ______ opinion.
   (a) ad hoc
   (b) adverse
   (c) qualified *
   (d) unqualified
   (e) mathematical

61. General ledger produces the sales journal and the ______ journal summarizing all transactions.
   (a) assets
   (b) liabilities
   (c) income
   (d) profit
   (e) cash receipts *

Decision Support Systems

62. ______ are commonly used interactive computer systems that support management decision-making activities.
   (a) MIS
   (b) DBMS
   (c) DAS
   (d) DSS *
   (e) EDP

63. With commonly used ______ users engage in what-if planning.
   (a) DSS *
   (b) EDP
   (c) MIS
   (d) DBMS
   (e) DAS

64. DSS is generally used in ______, that is, special purpose situations that may vary from day to day.
   (a) electronic worksheets
   (b) ad hoc inquiry and reporting *
   (c) sales journals
   (d) modeling
   (e) goal seeking

VisiCalc

65. VisiCalc is an example of a(n) ______.
   (a) calibration
   (b) ad hoc inquiry and reporting
   (c) electronic worksheet *
   (d) model
   (e) DAS

Modeling

66. The capability of a DSS system that enables managers to project the effects of their decisions using sophisticated mathematical techniques is called ______.
   (a) electronic worksheets
   (b) goal seeking
   (c) qualified opinions
   (d) sales analysis
   (e) modeling *

67. A ______ model might help executives determine what sales would be without the competitor's product on the market.
   (a) role
   (b) goal
   (c)DAS
   (d) mathemetical *
   (e) vendor

68. Changing a model to ensure that it is accurate is called ______.

(a) ad hoc inquiry and reporting
(b) calibration *
(c) goal seeking
(d) redundancy
(e) strategic planning

Case Study 11-3: An Example of What-If Planning

69. The ability of a DSS system that allows a manager to enter a certain result which the DSS then determines how to achieve is called ______.
(a) goal seeking *
(b) tactical planning
(c) calibration
(d) redundancy
(e) ad hoc inquiry and reporting

True/False

Data as a Business Resource

1. A business's data is one of its most precious resources. (T p274)

2. Most businesses view the collection of all their operating information as s single database. (F p274)

3. The process of keeping files current, complete, and up-to-date is called file maintenance. (T p274)

4. A general goal in designing systems is to avoid information redundancy. (F p275)

Database Management Systems

5. Databases are generally stored on secondary storage devices. (T p275)

6. A database may be used by only a single application system. (F p275)

7. The user of a database system must be familiar with the format in which the data is stored. (F p275)

8. The DBMS provides ready access to all users. (F p275)

9. Ensuring that what is supposed to be in the database is there, and what is not supposed to be there isn't, is called database integrity. (T p275)

10. The three common ways to structure a database are hierarchical database, network database, and relational database. (T p275)

Hierarchical Database

11. In a network database, data records are arranged in a strict parent/child relationship. (F p275)

12. In a hierarchical database, each parent record may have many children. (T p275)

13. Most hierarchical databases have many levels. (T p276)

14. Searching a hierarchical database is rapid and convenient as long as it is searched from the top down. (T p276)

15. The main advantage of a hierarchical database is that it can be searched conveniently in many different ways. (F p276)

Network Database

16. The network database is a more general structure than the hierarchical database. (T p277)

17. The network database offers the user multiple views. (T p277)

18. The network database structure is easier to maintain than the hierarchical database structure. (F p277)

19. The network database structure is the most user-friendly of the three common database structures. (F p278)

Relational Database

20. The relational database stores information as tables called relations. (T p278)

21. Relations in a relational database are sometimes called flat files. (T p278)

22. The columns of a relation in a relational database are very much the same as the records of a file. (F p278)

23. In a relational database, each row is called a domain. (F p278)

24. In a relational database, each column is called a tuple. (F p278)

25. In a relational database, fields are also called domains. (T p278)

26. In a relational database, combining two relations is called joining. (T p278)

27. In a relational database, selection is the process of extracting some of the columns from the relation and using them to form the columns of a new relation. (F p278)

28. One disadvantage of the relational database organization is that computers are not yet designed to manipulate relations efficiently. (T p278)

Case Study 11-1: A Simple Relational DBMS

29. In a relational database, one of the tuples is used as a key field. (F p278)

30. In a relational database, the select operation is used to extract records to form a new relation. (T p279)

31. In a relational database, a relation may be projected to remove an unwanted field. (T p279)

32. Each relation in a relational database may be used by only one system. (F p279)

Data Dictionary

33. In database systems, programs must contain detailed descriptions of all the data they process. (F p280)

34. In a database system, if the size or structure of a data item is altered, all the programs that reference that data must be modified. (T p280)

35. With database systems, the detailed descriptions of all the data items are kept in a centralized data dictionary. (T p280)

36. One advantage of the data dictionary in a database system is that it helps ensure standardization. (T p280)

Data Independence

37. The ability to modify the structure of the database without affecting existing programs that reference the database is called data independence. (T p281)

Management Information Systems

38. The key function of a manager is managing information. (F p281)

39. The primary goal of electronic data processing systems is to reduce clerical costs by computerizing the flow of day-to-day business transaction data. (T p281)

40. One reason why MIS was not successful in the 1960s was that business data was maintained primarily as files on disk. (F p281)

41. One reason why MIS was not successful in the 1960s was that data communications was much slower and more expensive than it is today. (T p281)

42. Today's view of an MIS is a set of related applications systems that share a company-wide database. (T p282)

43. Today, it is perfectly reasonable for a large company to maintain the vast majority of all its information on disk. (F p282)

44. As we move up the management hierarchy, the data provided to each level of management must be less detailed and more summarized. (T p282)

45. Medium-level managers tend to be involved with the day-to-day transactions of the business. (F p282)

46. Lower-level managers tend to deal with tactical planning. (F p282)

47. Upper-level managers tend to deal with strategic planning. (T p282)

48. Each level of management must receive information indicating how well all levels above it are performing. (F p282)

49. As we move up the management hierarchy, decisions tend to become less programmed. (T p283)

50. As we move down the management hierarchy, decisions tend to become less structured. (F p283)

Case Study 11-2: An MIS for FutureStore

51. Making repeated requests for payment on overdue accounts is called dunning. (T p284)

Order Entry System

52. An order entry system accepts customer orders and ensures that the orders are fulfilled. (T p284)

53. The accounts receivable system prints and mails customer invoices. (F p285)

54. The invoicing system monitors an invoice until it is paid. (F p285)

55. An order to deliver a certain amount of a product to a customer on the same day each week is called a standing order. (T p285)

Inventory Control System

56. One goal of an inventory control system is to maximize the amount of inventory on hand to ensure that customer requests can be fulfilled. (F p285)

57. Management uses the distribution-by-value report in an inventory control system to determine the most important items. (T p288)

58. The distribution-by-value report generally reveals that some small portion of the items account for the vast majority of the company's sales. (T p288)

Manufacturing Resources Planning System

59. MRP systems are especially important for companies that primarily offer services rather than products. (F p288)

60. MRP interfaces primarily with inventory control to determine what products need to be produced and when. (F p288)

61. MRP interfaces primarily with order entry to ensure that needed raw materials are available. (F p288)

62. The bill of materials is a key subsystem of MRP. (T p288)

63. The bill of materials is the bill the company pays for the materials it uses. (F p288)

64. The bill of materials shows the subcomponents of a component. (T p288)

65. Two reports produced by an MRP system are the material requirements and the capacity requirements. (T p289)

66. MRP interfaces primarily with order entry to obtain forecasts of needs. (F p289)

Billing System

67. When products are shipped to the customer, the billing system generates an invoice. (T p289)

68. The phrase "2% 10 days" appearing on an invoice means that 2% of the order must be paid for within 10 days of the invoice date. (F p289)

Accounts Receivable System

69. The accounts receivable system highlights "poor payers" so that credit and collection can focus its attention on these customers. (T p289)

70. The two major types of accounts payable systems are open item and balance forward. (F p289)

71. In an open item system, incoming cash is always applied to the oldest invoice. (F p289)

72. In a balance forward system, cash is applied to the outstanding balance. (T p289)

73. Some accounts receivable systems have forecasting features that can predict when customer payments are likely to arrive. (T p289)

74. The key report in the accounts receivable system is the aged trial balance. (T p290)

75. The ATB shows the age of each portion of the customer's balance. (T p290)

76. The accounts receivable system calculates finance changes on overdue amounts. (T p290)

77. Transactions that cannot be processed immediately are generally shown on suspense listings. (T p290)

Credit and Collection

78. Accounts receivable keeps track of promises to pay, refusals to pay, and the like. (F p291)

79. Credit and collection interfaces with invoicing to keep informed of all payments that have arrived so that it can promptly cease collection activity on accounts that have paid their bills. (F p291)

Payroll

80. Payroll is generally one of the first areas companies computerize. (T p291)

81. Three different types of payroll are salary, hourly, and incentive. (T p292)

82. Salaried employees normally receive extra pay for working overtime. (F p292)

83. It is not unusual for a company with its own computers to have its payroll processed by an outside organization. (T p292)

84. The payroll system produces a check register listing all checks paid. (T p292)

85. With direct deposit, employee's paychecks are mailed directly to their own banks. (F p294)

Accounts Payable System

86. One function of an accounts payable system is to help a company take advantage of discounts offered by suppliers for prompt payment. (T p294)

87. A good accounts payable system will pay bills as far in advance of their due dates as possible to take maximum advantage of discounts. (F p294)

88. All payment information in an accounts payable system is passed to the general ledger system. (T p294)

89. The cash requirements report is generated by the credit and collection system. (F p294)

90. The vendor master report is an essential report in the payroll system. (F p294)

General Ledger System

91. The general ledger system produces reports that summarize the financial health of the company. (T p294)

92. The chart of accounts is a listing of all the categories to which the general ledger system applies charges and payments. (T p294)

93. The two most important reports produced by the general ledger system are the profit and loss statement and the income statement. (F p294)

94. The difference between income and expenses is a company's assets. (F p294)

95. The difference between a company's assets and liabilities is its net worth. (T p294)

96. Outside auditing firms certify that income statements and balance sheets have been prepared "in accordance with generally accepted accounting principles." (T p294)

97. Upon completing an audit, the auditor will give a qualified opinion if the financial statements are in good order. (F p294)

98. Upon completing an audit, the auditor will give an adverse opinion if certain areas of the financial statements are weak. (F p294)

99. Upon completing an audit, the auditor will give an unqualified opinion if the financial statements as a whole are unacceptable. (F p294)

100. The sales journal and the cash receipts journal are produced by the sales analysis system and the credit and collection system, respectively. (F p294)

Decision Support Systems

101. What-if planning is a common activity with decision support systems. (T p296)

102. The DSSs of the 1980s represent a great leap forward from the DASs of the 1970s. (F p296)

103. "What would be the increase in sales if we double our advertising budget?" is a question likely to be asked of an MIS system. (F p296)

104. DSS has been used most productively in financial planning applications. (T p296)

105. DSS fully automates the decision process. (F p296)

106. Spoken English is probably the optimal form of executive-to-computer and computer-to-executive communications. (T p296)

107. MIS is generally used in ad hoc inquiring and reporting. (F p296)

VisiCalc

108. VisiCalc is an excellent example of a "what-if" DSS system. (T p298)

109. VisiCalc is an electronic worksheet system. (T p298)

Fill-in-the-Blanks

Data as a Business Resource

1. A **database** is ordinarily a large collection of related data items.

2. The process of keeping files correct, complete, and up-to-date is called **file maintenance** .

3. When the same data appears in several files we have what is called **redundancy** .

4. **Redundancy** is a useful security measure, but it is costly and could result in the use of out-of-date information.

Database Management Systems

5. When a user program inquires if a particular item is in the database, a **database management system (DBMS)** does the actual searching and retrieving.

6. The **DBMS** creates the database, keeps it up-to-date, and provides access to authorized users.

7. A DBMS ensures that what is supposed to be in the database is there, and what is not supposed to be there isn't; this is called database **integrity** .

8. The person who determines who may access the database, modify it, and add new relationships is the database **administrator** .

Hierarchical Database

9. In a **hierarchical** database, data records are arranged in a strict parent/child (boss/employee) relationship.

10. Searching a **hierarchical** database is rapid and convenient as long as it is searched from the top down.

Network Database

11. The **network** database is a more general structure than the hierarchical database; it can answer a wide variety of questions without having to provide duplicate data or to reorganize the database.

12. A network database uses **pointers** , special data items inside the computer that correspond to arrows in the diagram.

Relational Database

13. In a **relational** database, information is stored as tables.

14. The tables in a relational database are called **relations**.

15. Relations are often called **flat files** because the rows of the table really are very much the same as the records of a file.

16. Each row in a relation is called a record or a **tuple**.

17. Fields in a relational database are also called **domains**.

18. In a relational DBMS, two relations can be joined, a relation may be **projected**, and records may be selected.

Case Study 11-1: A Simple Relational DBMS

Data Dictionary

19. With database systems, the detailed descriptions of all the data items are kept in a centralized **data dictionary**.

20. The name, type, size, and other useful information about a data item would be found in the **data dictionary**.

Data Independence

21. The ability to modify the structure of the database without affecting existing programs that reference the database is called **data independence**.

Management Information Systems

22. Helping managers in their planning, organizing, directing, and controlling functions is the thrust of a **management information** system.

23. One of the earliest buzzwords in the field of computerized business systems was **electronic data processing** (EDP).

24. **Lower-level** managers tend to be involved with the day-to-day transactions of the business.

25. Medium-level managers tend to deal with short-term, **tactical** planning.

26. Upper-level managers tend to deal with long-term, **strategic** planning.

Case Study 11-2: An MIS for FutureStore

Order Entry System

27. An **order entry** system accepts customer orders and puts into motion the mechanisms to ensure that the orders are fulfilled.

28. Standing orders would be handled by the **order entry** system.

Inventory Control System

29. **Inventory** is the amount of each product a company keeps on hand to satisfy customer requests.

30. One goal of an **inventory control** system is to minimize the amount of inventory.

31. The **distribution-by-value** report lists every item in the inventory, the quantity sold for a particular period, and the dollar volume for each item.

Manufacturing Resources Planning System

32. A **manufacturing resources planning** system helps with the complex planning, resource allocation, and progress reporting functions associated with manufacturing a product.

33. A complete listing of all the components that go into manufacturing an item is called a **bill of materials**.

34. A tentative production schedule includes a material requirements report and a **capacity** requirements report.

Billing System

35. An invoice is produced by the **billing** system; it lists items shipped, quantities, and pricing for each item.

Accounts Receivable System

36. An **accounts receivable** system handles collections, and highlights poor payers.

37. The type of accounts receivable system that applies incoming cash to either a specific invoice or to the oldest invoice is called **open item**.

38. The accounts receivable system that applies cash to the outstanding balance is called the **balance forward** system.

39. An accounts receivable system produces the **aged trail balance (ATB)** report that shows the status of each customer including outstanding balance, current payments, and current purchases.

40. Transactions that must be handled manually are listed in a **suspense** listing.

Credit and Collection System

41. A **credit and collection** system maintains in-house credit records on all customers, and also uses credit information supplied by outside credit bureaus.

Payroll System

42. **Payroll** is one of the first areas companies computerize.

43. The **payroll** system is concerned with state and federal taxes, pension plan contributions, union dues, and life insurance premiums, as well as calculating gross pay.

44. In a **payroll** system, all checks paid are listed in a check register.

45. The ability of employees to have their pay placed into their bank accounts electronically by computer is called **direct deposit**.

Accounts Payable System

46. The job of the **accounts payable** system is to pay suppliers on time and take advantage of discounts for prompt payment.

47. The report produced by accounts payable that shows how much money is needed to handle forthcoming payments is the **cash requirements** report.

48. The report that lists all of the suppliers, their addresses, and their payment terms is the **vendor master** report.

General Ledger System

49. The **general ledger** system keeps track of financial records and produces reports summarizing the financial health of a company.

50. The **chart of accounts** is a listing of all the categories to which the general ledger system applies charges and payments.

51. The income statement, also called the **profit and loss** statement, summarizes by category all the income and expenses a business had for a reporting period.

52. The **balance sheet** summarizes assets and liabilities of a company.

53. What a company owns is called its **assets**.

54. **Liabilities** are what a company owes.

55. The difference between a company's assets and liabilities is the company's **net worth**.

56. If financial records are in good order, an auditor will issue an **unqualified** opinion.

57. If certain areas are weak, an auditor will issue a **qualified** opinion.

58. An **adverse** opinion is issued by an auditor if the financial statements are not acceptable.

59. The general ledger produces the sales journal and the **cash receipts** journal, listing and summarizing all financial transactions of a company.

Decision Support Systems

60. Interactive computer systems that support management decision-making activities are called **decision support systems (DSS)**.

61. Users engage in **what-if planning** in an interactive dialogue with the computer using a DSS.

62. DSS is generally used in **ad hoc inquiry and reporting**, special purpose situations that may vary from day to day.

VisiCalc

63. VisiCalc is an example of "what-if" DSS system; it is also called an **electronic**

worksheet .

Modeling

64. A DSS can help executives construct a **mathematical model** to estimate what the sales would be without the competitor's product on the market, what portion of the market the competitor is likely to capture, and, what the lost sales might be.

65. A model must constantly be adjusted to be sure it is an accurate representation of the business; these adjustments are called **calibration** .

Case Study 11-3: An Example of What-If Planning

66. **Goal seeking** is a feature of DSS systems; a goal is stated and the system performs certain manipulations to determine how to achieve that goal.

Chapter 12
Operating Systems

Multiple Choice

Systems Programs

1. The programs that control the operation of the computer are the ______ programs.
   (a) systems *
   (b) interactive
   (c) applications
   (d) batch
   (e) timesharing

2. The ______ programs instruct the computer how to solve specific user problems.
   (a) timesharing
   (b) batch
   (c) systems
   (d) applications *
   (e) interactive

3. The set of systems programs for a computer is called that computer's ______.
   (a) buffers
   (b) operating system *
   (c) channels
   (d) applications
   (e) systems

4. A ______ should have an understanding of the internal operation of computer hardware.
   (a) applications programmer
   (b) device driver
   (c) systems programmer *
   (d) host processor
   (e) quantum

5. Computer ______ are companies that supply general-purpose systems programs to user installationsto help manage their hardware.
   (a) resource managers
   (b) applications programmers
   (c) systems programmers
   (d) device drivers
   (e) vendors *

Applications Programs

6. ______ programs solve a user's data processing needs.
   (a) Systems
   (b) Applications *
   (c) Interactive
   (d) Timesharing
   (e) Batch

7. A ______ programmer must understand the information needs of an organization and how to use a computer to meet those needs.
   (a) resource
   (b) systems
   (c) applications *
   (d) batch
   (e) real-time

Operating Systems

8. The collection of systems programs that allows a computer to handle many functions previously performed by computer operators is called a(n) ______
   (a) operating system *
   (b) batch processing
   (c) double buffering
   (d) timesharing
   (e) device driver

Operating Systems Functions

9. Operating systems are primarily ______; they handle the hardware, programs and data.
   (a) real-time systems
   (b) relocatable programs
   (c) reentrant programs
   (d) resource managers *
   (e) interactive systems

10. The operating system's use of resources is called ______.
   (a) timesharing
   (b) time slicing
   (c) batch processing
   (d) buffering
   (e) overhead *

Batch Processing

11. User jobs that are submitted to a central computer installation in ______.
    (a) multiprogramming
    (b) batch processing *
    (c) real-time systems
    (d) channels
    (e) interactive systems

12. With ______, users do not interact with their executing programs.
    (a) real-time systems
    (b) interactive systems
    (c) batch processing *
    (d) multiprogramming
    (e) timesharing

Multiuser Operating Systems

13. With ______ operating systems, many users share the resources of a single computer system.
    (a) multiuser *
    (b) scheduling
    (c) real-time
    (d) round robin
    (e) fault tolerant

14. In a ______ operating system, a failure of one component will not necessarily make the entire system fail.
    (a) fault tolerant
    (b) round robin
    (c) scheduling
    (d) multiuser *
    (e) real-time

Multiprogramming

15. With the use of ______ operating systems, many jobs can be run "simultaneously."
    (a) flip-flop
    (b) multiprogramming *
    (c) buffering
    (d) real-time
    (e) file

16. In ______, a user can have several activities in progress at the same time.
    (a) double buffering
    (b) timesharing
    (c) thrashing
    (d) time slicing
    (e) multitasking *

Channels and Interrupts

17. Hardware ______ transfer data between primary storage and peripheral devices.
    (a) interrupts
    (b) filters
    (c) channels *
    (d) time slicers
    (e) buffer

18. Concurrent operations are called overlapped or ______ processing.
    (a) parallel *
    (b) real-time
    (c) flip-flop
    (d) quantum
    (e) time slice

19. A hardware signal that gains the attention of the processor is called a(n) ______.
    (a) channel
    (b) buffer
    (c) filter
    (d) interrupt *
    (e) time slice

Double Buffering

20. ______ helps take advantage of overlapped processing.
    (a) Filters
    (b) Double buffering *
    (c) Time slicing
    (d) Timesharing
    (e) Segmentation

21. The area of primary storage set aside to communicate between the processor and a channel is a ______.
    (a) filter
    (b) pipe

(c) buffer *
(d) page
(e) spool

22. Double buffering is also called ______ buffering.
(a) page
(b) spool
(c) filter
(d) segmentation
(e) flip-flop *

Storage Protection

23. In multiprogramming operating systems, ______ mechanisms prevent users from interferring with one another.
(a) storage protection *
(b) pipe
(c) filter
(d) segmentation
(e) device driver

24. With the use of ______, the highest and lowest main storage addresses of a program are loaded into the computer.
(a) encryption
(b) timesharing
(c) storage holes
(d) boundary registers *
(e) quantums

25. If a program tries to access information outside its range of addresses, the program will stop in a process called ______.
(a) error trapping *
(b) encryption
(c) quantums
(d) redirected outputs
(e) segmentation

Relocatable Programs

26. Programs that can be placed anywhere in primary storage are said to be ______ programs.
(a) absolute
(b) relocatable *
(c) reentrant
(d) interactive

(e) redundant

27. It is important that programs be ______ in a multiuser system because each time the program is run it may have to be placed in a different location.
(a) redundant
(b) interactive
(c) relocatable *
(d) reentrant
(e) absolute

Reentrant Programs

28. Programs that do not modify themselves are called ______ programs.
(a) absolute
(b) relocatable
(c) redundant
(d) reentrant *
(e) interactive

29. Several users can use ______ programs at the same time because these programs do not modify themselves.
(a) relocatable
(b) absolute
(c) redundant
(d) interactive
(e) reentrant *

Deadlock

30. In a multiuser system, when two users wish to use the same resource at the same time a ______ may occur.
(a) time slice
(b) deadlock *
(c) error trap
(d) interrupt
(e) quantum

31. Normally the only way out of a ______ is to terminate one or more of jobs to release a resource.
(a) deadlock *
(b) quantum
(c) error trap
(d) interrupt
(e) time slice

Time-Sharing

32. _____ operating systems multiprogram many interactive users.
   (a) Time-slice
   (b) Multitasking
   (c) Timesharing *
   (d) Round robin
   (e) Real-time

33. A common application on timesharing systems is _____.
   (a) time slicing
   (b) file extentions
   (c) storage holes
   (d) interactive program development *
   (e) dispatching

34. The maximum value set by a hardware timer is called a _____.
   (a) segment
   (b) quantum *
   (c) pipe
   (d) filter
   (e) limit

35. The process of dispatching the processor between users is called _____.
   (a) paging
   (b) segmentation
   (c) piping
   (d) voting
   (e) time slicing *

Real-Time Systems

36. The most important goal of a _____ system is responsiveness.
   (a) real-time *
   (b) timesharing
   (c) scheduling
   (d) dispatching
   (e) segmentation

37. An air traffic control system is an example of a _____ system.
   (a) segmentation
   (b) dispatching
   (c) real-time *

(d) scheduling
(e) timesharing

Scheduling

38. Deciding who gets which resources in a multiuser operating system is called ______
    (a) time slicing
    (b) timesharing
    (c) deadlock
    (d) scheduling *
    (e) dispatching

39. Assigning the processor to the next user in a multiuser system is called ______
    (a) time slicing
    (b) deadlock
    (c) timesharing
    (d) scheduling
    (e) dispatching *

40. Jobs are viewed as arranged in a circle and the processor moves from job to job in a ______ scheme.
    (a) round robin *
    (b) thrashing
    (c) print spooling
    (d) fragmentation
    (e) paging

41. In a ______ form of scheduling all users are treated in the same way.
    (a) paging
    (b) fragmentation
    (c) round robin *
    (d) thrashing
    (e) print spooling

42. When a steady stream of high-priority users prevent low-priority users from receiving service, ______ can occur.
    (a) thrashing
    (b) indefinite postponement *
    (c) paging
    (d) deadlock
    (e) fragmentation

Print Spooling

43. ______ is used in multiuser systems to allow users to run their jobs even when a printer is unavailable.
(a) Paging
(b) Segmentation
(c) Thrashing
(d) Print spooling *
(e) Voting scheme

44. Using ______ in a single-user system enables a program to run at the same time a file is being printed.
(a) thrashing
(b) voting scheme
(c) print spooling *
(d) paging
(e) segmentation

Virtual Storage

45. The distinction between primary and secondary storage disappears as far as the user is concerned in ______ storage.
(a) virtual *
(b) access control
(c) fault tolerant
(d) bootstrap
(e) real-time

46. With ______ storage the operating system creates the illusion of a single massive storage.
(a) parallel
(b) fault tolerant
(c) real-time
(d) access control
(e) virtual *

47. Computers can only execute instructions and reference data that are in its ______ storage.
(a) secondary
(b) primary *
(c) real-time
(d) virtual
(e) peripheral

Segmentation

48. The technique for implementing virtual storage in which all program and data pieces are different sizes is ______.

(a) paging
(b) fragmentation
(c) segmentation *
(d) thrashing
(e) print spooling

49. The technique for implementing virtual storage in which all program and data pieces are the same size is ______.
(a) print spooling
(b) thrashing
(c) segmentation
(d) paging *
(e) fragmentation

50. A ______ corresponds exactly to a meaningfully defined piece of a program.
(a) filter
(b) page
(c) storage hole
(d) pipe
(e) segment *

51. When many storage holes occur in primary storage we have what is called ______.
(a) fragmentation *
(b) segmentation
(c) paging
(d) thrashing
(e) print spooling

Paging

52. The use of ______ helps eliminate fragmentation.
(a) segmentation
(b) print spooling
(c) paging *
(d) thrashing
(e) filters

53. Data and program pieces that are the same size are called ______.
(a) filters
(b) pipes
(c) segments
(d) pages *
(e) storage holes

54. In paging, a new page is placed in a page ______ in primary storage.

(a) segment
(b) frame *
(c) hole
(d) filter
(e) list

55. Excessive paging causes what is known as ______.
(a) segmentation
(b) fragmentation
(c) buffering
(d) flip-flopping
(e) thrashing *

Multiprocessing

56. ______ operating systems manage more than one processor.
(a) Timesharing
(b) Multiprocessing *
(c) Real-time
(d) Uniprocessing
(e) Front-end

57. ______ is the amount of work a computer can process in a given amount of time.
(a) Real-time
(b) Paging
(c) Throughput *
(d) Access control
(e) Segmentation

58. When a system continues to operate even when components fail, it is said to be ______.
(a) fault tolerant *
(b) user-friendly
(c) redundant
(d) interactive
(e) absolute

59. When several processors work in parallel performing the same calculations in which the majority decision is chosen we have a ______.
(a) utility program
(b) bootstrap loader
(c) survivable system
(d) voting scheme *
(e) device driver

60. The more powerful processor in a multiprocessor organization is called a ______.

(a) bootstrap loader
(b) front-end processor
(c) device driver
(d) page frame
(e) host processor *

61. The less powerful processor in a multiprocessor organization that handles communications between terminals is the ______
(a) device driver
(b) computer networking
(c) front-end processor *
(d) bootstrap loader
(e) host processor

62. ______ involves computer systems that are connected to data communications links.
(a) Multiprocessing
(b) Computer networking *
(c) Multitasking
(d) Parallel processing
(e) Multiprogramming

63. With ______ operating systems, users can share resources with many separate computer systems.
(a) network *
(b) uniprocessor
(c) multiprocessor
(d) fault tolerant
(e) user-friendly

Security

64. The most common security scheme requires users to type in ______.
(a) pass keys
(b) encryption
(c) access control
(d) passwords *
(e) access rignts

65. ______ codes information stored in the computer.
(a) Access right
(b) Pass key
(c) Encryption *
(d) Access code
(e) Password

66. With a(n) ______ list, the people that are allowed to use a file are stored in the computer.
   (a) password
   (b) access control *
   (c) encryption
   (d) pass key
   (e) security

67. The operations a user can perform on a file are called that user's ______.
   (a) access controls
   (b) passwords
   (c) encryption
   (d) pass keys
   (e) access rights *

User-Friendliness

68. A computer's ease of use is called ______.
   (a) access control
   (b) help menus
   (c) user-friendliness *
   (d) access rights
   (e) applications programming interface

69. Options for users to take are listed in ______.
   (a) help menus *
   (b) random access memory
   (c) utility programs
   (d) positive confirmation
   (e) read-only memory

70. The use of ______ is helpful in preventing a user from accidentally committing serious errors.
   (a) help menus
   (b) positive confirmation *
   (c) ulitily programs
   (d) upwards mobility
   (e) device independence

71. The operating system provides ______ that help a user perform certain common tasks.
   (a) device independence
   (b) help menus
   (c) utility programs *
   (d) positive confirmation
   (e) upwards mobility

72. The operating can easily adapt to growth with ______.
    (a) positive confirmation
    (b) utility programs
    (c) help menus
    (d) upwards mobility *
    (e) device independence

73. With ______ users can write programs without considerations of how the data is organized.
    (a) help menus
    (b) upwards mobility
    (c) utility programs
    (d) positive confirmation
    (e) device independence *

Firmware

74. Read-only memory is an example of ______.
    (a) hardware
    (b) firmware *
    (c) device independence
    (d) software
    (e) systems call

75. Memory chips that have preloaded software that cannot be changed are called ______.
    (a) ROM *
    (b) RAM
    (c) BDOS
    (d) CPM
    (e) EPROM

76. A ______ is a program that reads a copy of the operating system from ROM to RAM.
    (a) device driver
    (b) systems call
    (c) bootstrap loader *
    (d) applications programming interface
    (e) pipeline

Case Study 12-1: CP/M

77. Vendors whose computers are capable of running CP/M software are said to be CP/M ______.
    (a) drivers
    (b) devices
    (c) filters

(d) compatible *
(e) pipes

78. The portion of CP/M that receives user requests is ______.
(a) BDOS
(b) CCP *
(c) BIOS
(d) DDT
(e) ASM

79. The ______ of CP/M controls the user's view of the computer.
(a) ASM
(b) DDT
(c) BDOS
(d) BIOS
(e) CCP *

80. The portion of CP/M that handles all disk input/output and manipulates disk files is ______.
(a) BDOS *
(b) CCP
(c) ASM
(d) BIOS
(e) DDT

81. With CP/M, disk files are normally scattered over the surfaces of disks in separate pieces called ______.
(a) device drivers
(b) filters
(c) extents *
(d) pipes
(e) logical devices

82. ______ is the component of CP/M that handles input/output to all devices other than disks.
(a) ASM
(b) BIOS *
(c) DDT
(d) CCP
(e) BDOS

83. The separate programs contained in the BIOS component of CP/M are called ______.
(a) filters
(b) pipes
(c) logical devices
(d) device drivers *
(e) extents

84. All CP/M files have ______ that identify the type of file.
   (a) device drivers
   (b) filters
   (c) pipes
   (d) logical devices
   (e) file extensions *

85. ______ help the user and CP/M manage files and avoid making certain types of errors.
   (a) Logical devices
   (b) Pipes
   (c) File extensions *
   (d) Filters
   (e) Device drivers

86. Applications programs written to run under CP/M see only ______
   (a) logical devices *
   (b) filters
   (c) device drivers
   (d) pipes
   (e) file extensions

87. An applications program on CP/M executes a ______ that actually performs the required function.
   (a) file extension
   (b) systems call *
   (c) filter
   (d) device driver
   (e) logical device

88. An ______ is a view of the computer that looks the same on all CP/M computers regardless of the physical differences between the machines.
   (a) systems call
   (b) logical device
   (c) file extension
   (d) applications programming interface *
   (e) standardized operating system

89. One of the benefits of CP/M is that it is a ______
   (a) applications programming interface
   (b) file extension
   (c) standardized operating system *
   (d) systems call
   (e) logical device

Case Study 12-2: UNIX, XENIX, MS-DOS

UNIX:

90. ______ was developed at Bell Laboratories for use with 16-bit minicomputers.
    (a) MS-DOS
    (b) XENIX
    (c) CP/NET
    (d) CP/M
    (e) UNIX *

91. A ______ in UNIX is a program that reads a stream of inputs, processes them, and produces a stream of outputs.
    (a) pipe
    (b) systems call
    (c) filter *
    (d) extent
    (e) device driver

92. In UNIX, a ______ is a communication path between programs.
    (a) pipe *
    (b) device driver
    (c) systems call
    (d) extent
    (e) filter

93. In UNIX filters and pipes may be connected to form a ______.
    (a) systems call
    (b) pipeline *
    (c) network
    (d) device driver
    (e) file extension

XENIX:

94. A licensed version of UNIX that was designed by Microsoft is ______.
    (a) CP/M
    (b) MS-DOS
    (c) PC-DOS
    (d) XENIX *
    (e) MP/M-II

95. The ______ version of UNIX is considered to be more user-friendly.
    (a) PC-DOS
    (b) MP/M-II
    (c) MS-DOS

(d) CP/M
(e) XENIX *

MS-DOS:

96. ______ was developed by Microsoft for the IBM Personal Computer.
(a) XENIX
(b) CP/M
(c) MS-DOS *
(d) MP/M-II
(e) CP/NET

97. An optional operating system for the IBM Personal Computer is ______
(a) CP/M
(b) MP/M-II
(c) XENIX
(d) CP/M-86 *
(e) MP/M-92

98. The IBM version of MS-DOS is called ______
(a) PC-DOS *
(b) CPM-DOS
(c) MC-DOS
(d) IBM-DOS
(e) Blue-DOS

True/False

Operating Systems

1. Systems programs control the operation of the computer. (T p308)

2. Applications programs instruct the computer how to solve specific user problems. (T p308)

3. The set of applications programs on a computer is called that computer's operating system. (F p308)

Systems Programs

4. One goal of systems programs is to make compiex computer hardwrare more user-friendly. (T p308)

5. Applications programming requires a real understanding of the internal operation of computer hardware. (F p308)

Applications Programs

6. Systems programs solve a user's data processing needs. (F p308)

7. Applications programmers generally must know more about the workings of a company than systems programmers. (T p308)

8. The "view" of the computer system seen by an applications programmer is provided by the hardware. (F p308)

9. The view of the computer system seen by a systems programmer is provided by the operating system. (F p308)

10. The view of the computer system seen by the user is provided by the applications programs. (T p308)

Operating Systems

11. In the 1950s, the cost of computer hardware was much greater than the cost of computer operators; today the opposite is true. (T p308)

12. Operating systems first appeared in the late 1960s. (F p308)

Operating Systems Functions

13. Operating systems are primarily resource managers. (T p309)

14. The operating system's use of resources is called overhead. (T p309)

Batch Processing

15. Removing a job from the computer system is called "ripping off." (F p310)

16. In batch processing, users generally do not interact with their executing programs. (T p310)

Multiuser Operating Systems

17. One advantage of multiuser computer systems is that if the system fails, most of the users can continue operating. (F p310)

Multiprogramming

18. Multitasking is a special case of multiprogramming in which each user may have multiple activities in progress at once. (T p310)

Channels and Interrupts

19. Channels can operate in parallel with the CPU. (T p310)

20. When the CPU is doing calculations, the channels must wait. (F p310)

21. To perform an input or output operation, the processor executes an instruction telling the channel to begin the operation; the processor may then perform calculations for other users. (T p310)

22. When an input or output operation finishes, the processor generates a hardware signal called an interrupt to gain the attention of the channel. (F p310)

Double Buffering

23. With double buffering, an area of secondary storage, called a buffer, is set aside for communication between the processor and a channel. (F p310)

24. With flip-flop buffering, while the channel is transmitting from one buffer, the CPU can fill the other buffer. (T p310)

Storage Protection

25. One popular scheme for enforcing storage protection is the use of boundary registers. (T p311)

26. When a user program refers to a storage address outside the range specified in the boundary registers, the hardware stops the reference before any damage can be done; this is an example of error tapping. (T p311)

Relocatable Programs

27. Programs that can only run if put in certain fixed locations are called absolute. (T p311)

28. Programs that may be placed anywhere in primary storage are said to be redundant. (F p312)

29. Making user programs absolute is essential in multiuser systems. (F p312)

Reentrant Programs

30. Programs that do not change while being executed are called reentrant programs. (T p312)

31. Several users may share and execute the same copy of a program simultaneously if that program is relocatable. (F p312)

Deadlock

32. Deadlocks of various kinds can occur in almost every computer system. (T p313)

33. When deadlock does occur, normally the only way out is to terminate all of the deadlock jobs. (F p313)

Timesharing

34. Timesharing operating systems multiprogram many interactive users. (T p313)

35. Interactive program development is commonly performed on batch processing systems. (F p313)

36. One key to guaranteeing acceptable response times in timesharing systems is using a hardware timer. (T p313)

37. The quantum in a timesharing system is ordinarily set to a few seconds. (F p313)

38. When the quantum expires, an interrupt is generated to get the attention of the processor. (T p313)

39. The scheme of offering brief slices of time to each user in a timesharing system is called the salami method. (F p313)

Real-Time Systems

40. The most important goal in real-time systems is servicing as many users as possible. (F p313)

Scheduling

41. The special case of scheduling the processor and assigning it to the next user is called dispatching. (T p313)

42. Round robin is an effective scheduling scheme for giving priority to certain users. (F p313)

43. Round robin is caosidered to be a fair scheduling scheme. (T p313)

44. A system that does enforce priorities could cause low-priority users to be indefinitely postponed. (T p313)

45. Some systems avoid indefinite postponement by gradually decreasing the priorities of low-priority users the longer they wait for service. (F p314)

Print Spooling

46. Print spooling allows users to run their programs even if a printer is not available. (T p314)

47. Whenever a user program tries to write to the printer in a system with print spooling, the spooling mechanism of the operating system intercepts the request and writes the data to disk; when the printer becomes available, the disk file may then be printed. (T p314)

48. Because printer output is so much faster than disk output, spooling makes it possible for programs that use disks (and other relatively slow devices) to run much faster. (F p314)

49. Print spooling is not useful on single-user systems. (F p314)

50. Print spooling that allows a user to run a program while a file is being printed is the most common form of multitasking. (T p314)

Virtual Storage

51. Secondary storage devices like disk and tape remain expensive compared to primary storage. (F p314)

52. In virtual storage systems, the distinction between primary and secondary storage disappears, at least as far as the user is concerned. (T p314)

53. It is common for a computer's primary storage to be 100 to 1000 times greater than its

virtual storage. (F p314)

54. One disadvantage of virtual storage systems is that users may tend to write programs that are larger than necessary. (T p314)

55. A computer can only execute instructions and reference data that are in its secondary storage. (F p314)

Segmentation

56. With the segmentation technique for implementing virtual storage, program and data pieces may be of different sizes. (T p314)

57. In the paging method for implementating virtual storage, program and data pieces are all the same size. (T p314)

58. With paging, each piece being shuttled back and forth normally represents a whole program, a whole data structure, or a whole procedure. (F p314)

59. A segment corresponds exactly to a meaningfully defined piece of a program. (T p314)

60. Segmentation tends to leave storage with many small unusable free areas that vary in size; this is called fragmentation. (T p315)

61. Reclaiming usable space in a segmentation system is called garbage collection. (T p315)

62. Garbage collection may be accomplished in a segmentation system be compacting all the occupied areas to one end of storage so that a single large usuable area remains. (T p315)

Paging

63. One advantage of paging over segmentation is that paging eliminates fragmentation. (F p315)

64. In a paging system, When all the page frames are full, the system is deadlocked. (F p315)

65. In order for a program to run efficiently in a paged system, its most active pages must be kept in secondary storage. (F p316)

66. Thrashing occurs when there is not enough room in primary storage to hold all of a program's pages. (F p316)

67. Some virtual storage systems combine paging and segmentation. (T p316)

Multiprocessing

68. Multiprocessing operating systems manage computers that have more than one processor. (T p316)

69. Uniprocessor systems are generally more reliable tham multiprocessor systems because the greater the number of processors, the greater the chance of failure. (F p316)

70. Fault tolerant or survivable systems are frequently implemented with multiprocessors. (T p316)

71. The reliability of a uniprocessor system may be increased by employing a voting scheme. (F p316)

72. One popular multiprocessing scheme uses a host processor to relieve the burden on a front-end processor. (F p316)

Security

73. One common technique for "cracking" computer systems is to try all possible passwords until the correct one is found. (T p316)

74. If a user unsuccessfully attempts to log-in, perhaps because of a typing error, the operating system should immediately refuse to accept futher communications from that terminal. (F p316)

75. If files are encrypted, then even if someone gains access to someone else's files, the information will be useless. (T p316)

76. A person who has been granted "R + E" access to a file may not make a copy of it. (F p316)

77. Write access allows the user to destroy a file at will. (T p317)

User-Friendliness

78. The user-friendly technique of positive confirmation might prevent a user from accidentally deleting a critical file. (T p318)

79. An accounts receivable system is an example of a utility program. (F p318)

80. A sort program or a merge program would generally be considered a utility program. (T p318)

81. Upward mobility means that operating systems designers tend to get frequent raises. (F p318)

82. Device independence is an operating systems feature that allows users to write programs independently of considerations of how data is organized on particular devices. (T p318)

Firmware

83. ROMs are generally considered to be software. (F p319)

84. The computer cannot modify the program in ROM. (T p319)

85. Conventional computer memory is called RAM for remote access memory. (F p319)

86. Operating systems are frequently placed in RAM for security. (F p319)

87. One popular application of ROM is to contain a special program called a bootstrap loader. (T p319)

88. The key difference between PROM and ROM is that PROM can be erased. (F p319)

89. An EPROM can be reprogrammed as needed. (T p319)

Case Study 12-1: CP/M

90. CP/M is considered by many personal computer users to be the standard operating system of the 32-bit microcomputer industry. (F p319)

91. A computer is said to be CP/M compatible if it is able to run CP/M. (T p319)

History of CP/M

92. IBM produced the first microprocessor chip in 1970. (F p319)

93. The first microcomputer was the Apple. (F p319)

94. Intel introduced the floppy disk, a highly reliable and relatively fast secondary storage device. (F p319)

95. Gary Kildall formed Microsoft in 1976 to develop, refine, and market CP/M. (F p319)

The Console Command Processor

96. CCP is the component of CP/M that receives user requests, and calls upon other components of CP/M to carry out the chores needed to service these commands. (T p319)

The Basic Disk Operating System

97. BDOS is the CP/M component that handles all disk input/output and manipulates disk files. (T p320)

98. CP/M disk files are normally scattered over the surface of disks in separate pieces called extents. (T p320)

99. BIOS and CCP make up the fixed portion of CP/M. (F p320)

The Basic Input/Output System

100. BIOS is the CP/M component that handles input/output to all devices other than disks. (T p320)

101. Reading a character from the console keyboard in CP/M is accomplished by CCP. (F p320)

102. Writing a character to the printer in CP/M is accomplished by BIOS. (T p320)

103. BIOS is the customized part of CP/M. (T p320)

Some Important CP/M Commands (Fig. 2, p321)

104. The CP/M command DDT displays the status of devices and files. (F p321)

105. The CP/M command DIR displays the names of files on a disk. (T p321)

106. The CP/M command PIP transfers files between devices. (T p321)

107. The CP/M command ED creates a new edition of the operating system. (F p321)

108. The CP/M command LOAD loads a program from disk into main storage. (T p321)

File Extensions

109. All file names in CP/M have a file extension, a period followed by the three letters CPM. (F p321)

110. The file extension .BAS in CP/M indicates a BASIC source program. (T p321)

111. The file extension .ASM in CP/M indicates an assembler source program. (T p322)

112. The file extension .MAC in CP/M indicates a Macintosh conversion program. (F p322)

Logical Devices

113. Applications programs written to run under CP/M see only physical devices. (F p322)

114. The CP/M applications program that wants to communicate with the user does so through "CON:". (T p322)

115. The CP/M applications program that wants to print lines on a hard-copy device does so through "PRT:". (F p322)

116. "RDR:" and "PUN:" are used by CP/M applications programs to read and write from a secondary storage device such as a tape cassette, a floppy disk, or a Winchester disk. (T p322)

Systems Calls

117. All CP/M machines are physically identical but they may have certain logical differences. (F p323)

118. Applications software that runs on one CP/M computer usually requires minor modifications to run on any other. (F p323)

119. One of the most important reasons users buy CP/M is to gain access to the huge amount of applications software available for CP/M systems. (T p323)

120. CP/M is a widely standardized operating system. (T p323)

The CP/M Family of Operating Systems

121. CP/M operating systems with "-86" in their names are designed to run on 16-bit microcomputers based on Intel's 8086 and 8088 microprocessors. (T 323)

122. Concurrent CP/M-86 is a multiuser operating system. (F p324)

123. CP/M-86 is a multitasking operating system. (F p324)

124. CP/M Plus was designed to be a more user-friendly operating system. (T p324)

Case Study 12-2: UNIX, XENIX, MS-DOS

125. UNIX was developed at Bell Laboratories in the late 1960s and early 1970s. (T p324)

126. UNIX is essentially a single-user, multitasking system. (F p324)

127. A UNIX filter is a program that reads a stream of inputs, processes them in some manner, and produces a stream of outputs. (T p 324)

128. A UNIX pipe is a communications path between programs. (T p325)

129. UNIX pipelines allow programs to be composed by combining existing pieces. (T p325)

130. Many companies market UNIX pipes. (F p325)

131. UNIX is an extremely user-friendly system designed primarily for unsophisticated users. (F p325)

132. UNIX excels at being a "programmer-friendly" system. (T p325)

133. XENIX is a licensed version of UNIX designed espically for 8-bit microcomputers. (F p325)

MS-DOS

134. MS-DOS was developed by IBM for the IBM PC. (F p325)

135. MS-DOS is expected to evolve towards CP/M. (F p325)

136. CP/M is not available for the IBM PC. (F p325)

Fill-in-the-Blanks

1. **Systems** programs control the operation of the computer.

2. **Applications** programs instruct the computer how to solve specific user problems.

3. The set of systems programs on a computer is often called that computer's **operating system** .

Systems Programs

4. **Systems** programs make complex computer hardware more user-friendly.

5. A **systems** programmer must have an understanding of the internal operation of computer hardware.

6. Companies the build computer hardware are called computer **vendors** .

7. Computer **vendors** generally supply general-purpose systems programs to user installations that use their hardware.

Applications Programs

8. **Applications** programs solve a user's data processing needs.

9. **Applications** programmers must understand the information needs of their organization and how to use computers to meet those needs.

10. **Systems** programmers see the hardware and write programs that depend on the specifics of the hardware.

11. **Applications** programmers see the operating system and the facilities it offers.

Operating Systems

12. **Operating systems** are collections of systems programs that allow computer systems to handle many of the functions previously performed by computer operators.

Operating Systems Functions

13. Operating systems are primarly **resource managers** ; they manage computer hardware, programs, and data.

14. The operating system's use of resources is called **overhead** .

Batch Processing

15. In **batch processing** , users submit their jobs to a central computer installation.

Multiuser Operating Systems

16. **Multiuser** operating systems allow many users to share the resources of a single computer system.

Multiprogramming

17. **Multiprogramming** operating systems allow many jobs to be run simultaneously.

18. **Multitasking** is a special case of multiprogramming in which each user may have multiple activities in progress at once.

Channels and Interrupts

19. Hardware **channels** transfer data between primary storage and peripheral devices.

20. When the CPU is doing calculations, channels can be transferring data; this is called overlapped processing or **parallel processing** .

21. The channel generates a hardware signal called an **interrupt** to gain the attention of the processor.

Double Buffering

22. **Double** buffering helps take advantage of overlapped processing.

23. An area of primary storage, called a **buffer** , is set aside for communication between the processor and a channel.

24. Double buffering is also called **flip-flop** buffering.

Storage Protection

25. **Storage protection** mechanisms prevent users from interfering with one another or with the operating system.

26. The storage protection scheme that uses hardware registers loaded with the highest and lowest main storage addresses of a user is called **boundary registers** .

27. **Error trapping** stops the hardware before any damage can be done to another's users programs or data.

Relocatable Programs

28. Programs that can only run if put in certain fixed locations are called **absolute** .

29. Programs that may be placed anywhere in primary storage are said to be **relocatable** programs.

Reentrant Programs

30. Programs that do not modify themselves are **reentrant** programs.

Deadlock

31. Deadlock is sometimes called **gridlock** .

Timesharing

32. **Interactive** users access the computer system through terminals.

33. **Timesharing** operating systems mulitprogram many interactive users.

34. The common application of timesharing systems that allows a programmer to type in program statements, translate and test the program, make corrections, and continue to process the program until it executes properly is called **interactive program** development.

35. When using a hardware timer in timesharing systems, the maximum value the timer is set to is called the **quantum** .

36. The scheme of offering only brief "bursts" of time to each user on a timesharing system is called **time slicing** .

Real-Time Systems

37. **Real-time** systems are computer systems that respond immediately to various occurrences.

38. The most important goal of a real-time system is its **responsiveness** .

Scheduling

39. In multiuser operating systems, deciding which user gets which resources and when is called **scheduling** .

40. The special case of scheduling the processor and assigning it to the next user is called **dispatching** .

41. In timesharing systems, users can be dispatched in **round robin** fashion in which the jobs are viewed as arranged in a circle and the processor moves from job to job in order around the circle.

42. **Indefinite postponement** occurs when a steady stream of high-priority users prevents low-priority users from receiving any service at all.

Print Spooling

43. With **print spooling** users can run their programs even if the printer is not available.

Virtual Storage

44. In **virtual storage** systems the distinction between primary and secondary storage disappears, at least as far as the user is concerned.

Segmentation

45. In the **segmentation** technique for implementing virtual storage, the program or data pieces may be of different sizes.

46. In segmentation, a **segment** corresponds exactly to a meaningfully defined piece of a program.

47. As segments are loaded and unloaded from primary storage, the storage **holes** that are left may vary in size.

48. With segmentation, many small holes dot primary storage, and these may be so small as to be unusable, causing a serious problem called **fragmentation** .

49. Reclaiming usable space by compacting all the occupied areas to one end of storage so that a single large usable area remains is called **garbage collection** .

Paging

50. The technique for implementing virtual storage where all program and data pieces are the same size is called **paging**.

51. In paging, when a new page is brought into main storage, the page may be placed in the space, called a **page frame**, vacated by any previous page.

52. All program and data items in paging are the same size called the **page size**.

53. In the paging technique for implementing virtual storage, excessive paging activity is called **thrashing**.

Multiprocessing

54. **Multiprocessing** operating systems manage computers that have more than one processor.

55. The amount of work a computer can process in a given amount of time is referred to as **throughput**.

56. A computer that only has a single processor is referred as a **uniprocessor** system.

57. Multiprocessors enable the development of highly reliable systems that keep working even when some components fail; these are sometimes called **fault tolerant** or survivable systems.

58. With a **voting scheme**, several computers perform the same calculations and the decision of the majority is followed.

59. In some multiprocessor organizations, the more powerful processor called the **host** processor, handles the heavy duty calculations.

60. In some multiprocessor organizations, the **front-end** processor handles the communications between the terminals and the host processor.

61. Closely related to multiprocessing is **computer networking** in which many computer systems are connected in a data communications network.

62. A **network** operating system allows users to share the resources of many separate computer systems.

Security

63. Secret **passwords** are used to gain access to multiuser computer systems, and to access files.

64. **Encryption** , the encoding of information, may be used in operating systems for additional security.

65. An **access control** list indicates the users who may access a file, and the operations each user may perform on the file.

66. The operations that each user may perform on a file are called that user's **access rights** .

User-Friendliness

67. The term used to describe a computer's ease of use is **user-friendliness** .

68. A feature of an operating system that helps users from accidentally committing a serious error is **positive** confirmation.

69. **Utility** programs are provided by the operating system to help the user accomplish certain common tasks.

70. **Upwards mobility** means that the operating system adapts easily to the growth of a computer system.

71. **Device independence** allows users to write programs independently of considerations of how data is organized on particular devices.

Firmware

72. Read-only memory is an example of what is called **firmware** .

73. **ROMs** are hardware memory chips that have been preloaded with software.

74. The computer cannot modify a program in a **ROM** chip.

75. The conventional memories that hold user programs are called **RAMs** .

76. One popular application of ROM is to contain a special program called a **bootstrap** loader.

77. A user programs a **PROM** for a particular application, but it can be programmed only once.

78. An **EPROM** can be reprogrammed as needed.

Case Study 12-1: CP/M

79. One of the most widely used operating systems in the world is **CP/M** originally designed for 8-bit microcomputers.

History of CP/M

80. CP/M was developed by **Gary Kildall** .

The Console Command Processor

81. The portion of CP/M that receives user requests and calls upon the other components of CP/M to carry out the chores needed to service these commands is the **Console Command Processor (CCP)** .

The Basic Disk Operating System

82. The **Basic Disk Operating System (BDOS)** is the component of CP/M that handles all disk input/output and manipulates disk files.

83. Disk files, in CP/M, are normally scattered over the surfaces of disks in separate areas called **extents** .

84. **BDOS** and CCP make up the fixed part of CP/M.

The Basic Input Output System

85. The component of CP/M that handles input/output to all devices other than disk is the **Basic Input Output System (BIOS)** .

86. The BIOS of CP/M contains many separate programs called **device drivers** to handle each type of device.

File Extensions

87. All file names in CP/M have a **file extension** , a period followed by a three-letter code that helps identify the type of file.

Logical Devices

88. Applications programs written to run under CP/M see only **logical** devices.

89. CP/M allows users to **redirect** outputs to different devices.

System Calls

90. Applications programs in CP/M do not perform a request directly; rather they execute a **system call**; this calls upon the appropriate portion of CP/M to perform the actual chore.

91. System calls and logical devices create an **applications** programming interface--a view of the computer that looks the same on all CP/M computers regardless of the physical differences between the machines.

92. One of the benefits of any widely **standardized** operating system is that hardware vendors who build their machines to run CP/M can attract many more buyers.

The CP/M Family of Operating Systems

93. Versions of CP/M with **-86** in their names are designed to run on 16-bit microcomputers.

Case Study 12-2: UNIX, XENIX, MS-DOS

UNIX

94. **UNIX** is an operating system developed by Bell Laboratories in the late 1960s and early 1970s.

95. In the UNIX operating system, a **filter** is a program that reads a stream of inputs, processes them in some manner, and produces a stream of outputs.

96. A **pipe** is a communication path between programs in the UNIX operating system.

97. In the UNIX operating system, filters and pipes may be connected to form a **pipeline** in which inputs enter and go through a series of processing filters, and from which outputs emerge.

XENIX

98. **XENIX** is a licensed version of the UNIX system designed by Microsoft espicially for 16-bit microcomputers.

MS-DOS

99. The operating system developed by Microsoft for the IBM Personal Computer is **MS-DOS** , which IBM calls PC-DOS.

Chapter 13
Personal Computing

Multiple Choice

What Is a Personal Computer?

1. A personal computer is a complete computer system that is small enough to fit on a ______
   (a) suitcase
   (b) briefcase
   (c) desktop *
   (d) den
   (e) playroom

2. Which of the following is not a characteristic of a personal computer? ______
   (a) requires special lessons *
   (b) user-friendliness
   (c) reliable
   (d) serviceable
   (e) able to run large amount of software

A Brief History of Personal Computing

3. The key development that made personal computers possible was the invention of the ______
   (a) mouse
   (b) icon
   (c) keyboard
   (d) floppy disk
   (e) microprocessor *

4. The first ______ were not very powerful and were expensive.
   (a) plotters
   (b) microprocessors *
   (c) modems
   (d) printers
   (e) sturdy disks

5. Steve Jobs and Steve Wozniak demonstrated their first computer to ______
   (a) Atari
   (b) Hewlett-Packard

(c) IBM
(d) Home Brew Computer Club *
(e) Tandy Corporation

6. The ______ is marketed through the Radio Shack retail stores.
(a) TRS-80 *
(b) PET
(c) IBM PC
(d) Commodore
(e) Atari

7. In 1981 ______ finally entered the personal computer market.
(a) Atari
(b) Tandy Corporation
(c) IBM *
(d) Hewlett-Packard
(e) Commodore

Case Study 13-1: The IBM Personal Computer

8. The IBM PC hardware is ______ designed so that various pieces can be combined to meet different needs.
(a) on-line
(b) modularly *
(c) independently
(d) interactively
(e) individually

9. The ______ unit of the IBM PC houses the microprocessor.
(a) keyboard
(b) display
(c) printer
(d) system *
(e) expansion

10. The RAM and ROM of the IBM PC are located in the ______ unit.
(a) printer
(b) expansion
(c) display
(d) keyboard
(e) system *

11. The ______ allow the user to add memory, diskette drives and communication devices to the IBM PC.
(a) megabytes

(b) systems boards
(c) exansion slots *
(d) systems units
(e) display units

12. The ______ feedback feature of the IBM PC makes a sound when a key is struck.
(a) tactile
(b) audio *
(c) modem
(d) monochrome
(e) simulation

13. The IBM printer can set any dot on the page on or off because it uses ______.
(a) monochrome display
(b) expert systems
(c) pull-down menus
(d) APA graphics *
(e) plotters

Case Study 13-2: The Apple Macintosh

14. The ______ reestablished Apple as a leader in the personal computer market.
(a) Apple IIe
(b) Apple II
(c) Lisa
(d) Star
(e) Macintosh *

15. The ______ is used with the Macintosh to carry on a dialogue between the user and the computer.
(a) mouse *
(b) joystick
(c) touch sensor
(d) APA graphics
(e) sturdy disk

16. A(n) ______ describes the operation to be performed on the Macintosh.
(a) sturdy disk
(b) icon *
(c) mouse
(d) joystick
(e) clicker

17. The user presses the button on the mouse to indicate what operation is to be performed, this is called ______.

(a) bidirectional printing
(b) displaying
(c) clicking *
(d) plotting
(e) telecomputing

18. The use of a mouse, icons, and pointing was first introduced by ______ for the Star workstations.
(a) Tandy Corporation
(b) IBM
(c) Comodore
(d) Xerox *
(e) Atari

19. The command on the Macintosh that removes a piece of text from a document is ______.
(a) Paste
(b) Print
(c) Point
(d) Edit
(e) Cut *

20. The ______ command on the Macintosh places a piece of text into another document.
(a) Edit
(b) Paste *
(c) Print
(d) Point
(e) Cut

21. The ______ interface on the Macintosh allows a modem to be connected.
(a) printer
(b) Imagewriter
(c) RS-232 *
(d) clicking
(e) audio

Portable Computers

22. A(n) ______ is a computer that can be easily carried by users wherever they go.
(a) portable computer *
(b) word processor
(c) expert system
(d) videodisk
(e) on-line database

Choosing a Personal Computer

23. The best place to purchase a computer is at a ______.
    (a) department store
    (b) mail order firm
    (c) wholesaler
    (d) computer store *
    (e) vendor supply ad

24. As a rule you should consider at least ______ alternatives before making a decision to purchase a personal computer.
    (a) 5
    (b) 3 *
    (c) 2
    (d) 4
    (e) 7

Personal Computer Data Communications

25. The two primary components that enable a personal computer to communicate over the standard phone lines are an RS-232 interface and a ______.
    (a) imagewriter
    (b) teletext
    (c) database manager
    (d) expert system
    (e) modem *

26. The instructions that enable a personal computer to communicate are found in a ______.
    (a) database manager
    (b) modem
    (c) communications software package *
    (d) teletext
    (e) information provider

Applications

27. The application for personal computers most used by students and business people is ______
    (a) word processing *
    (b) database management
    (c) computer assisted instruction
    (d) data communications
    (e) electronic spreadsheets

28. The application of a personal computer for organizing household budgets, or managing

membership lists is called ______.
(a) electronic spreadsheets
(b) word processing
(c) data communications
(d) database management *
(e) computer assisted instruction

29. The ability of working at home using a computer terminal is called ______
(a) data communications
(b) teletext
(c) viewdata
(d) on-line databases
(e) telecommuting *

Personal Computing Software

30. The software package that controls the computer is the ______
(a) database management system
(b) operating system *
(c) word processing system
(d) electronic spreadsheet
(e) programming language

31. ______ software is used primarly for financial planning applications.
(a) Word processing
(b) Operating system
(c) Database management
(d) Electronic spreadsheet *
(e) Data communications

32. With ______ software, a user can prepare charts.
(a) graphics *
(b) expert system
(c) word processing
(d) electronic spreadsheet
(e) database management

33. Users can obtain advice in such fields as law and first aid by using ______ systems.
(a) operating
(b) database management
(c) expert *
(d) graphics
(e) data communications

Education

34. Which of the following should not be associated with the use of computers in education? ______
 (a) Fun to use
 (b) Confusing *
 (c) Learn at home
 (d) Self paced
 (e) Customized

35. One of the tools that was replaced by the calculator was the ______.
 (a) ruler
 (b) pencil
 (c) slide rule *
 (d) supercomputer
 (e) Jacquard loom

36. In recent years ______ have become popular where students mix computer courses with athletic activities.
 (a) computer camps *
 (b) computer schools
 (c) computer clubs
 (d) computer societies
 (e) computer programs

37. The use of computers to help the learning process is called ______.
 (a) computer courses
 (b) calculating
 (c) expert systems
 (d) viewdata
 (e) computer assisted instruction *

38. Computers connected to ______ make especially effective interactive teaching machines.
 (a) printers
 (b) graphics terminals
 (c) viewdata systems
 (d) laser videodisks *
 (e) viewtron systems

Information Providers

39. Commercial enterprises that distribute data electronically are called ______.
 (a) interactive teaching systems
 (b) information providers *
 (c) computer assisted instructors

(d) videodisk machines
(e) expert systems

Teletext

40. The information service that broadcasts in only one direction is ______.
    (a) viewdata
    (b) on-line databases
    (c) teletext *
    (d) videotext
    (e) full definition video images

41. Teletext information is transmitted via ______.
    (a) vertical blanking intervals *
    (b) alphamosaic graphics
    (c) full definition video
    (d) alphageometric graphics
    (e) closed captioning

42. The first widespread use of teletext was the ______ of TV programs for the deaf.
    (a) alphageometric graphics
    (b) full definition video
    (c) vertical blanking intervals
    (d) alphamosaic graphics
    (e) closed captioning *

Viewdata/Videotex

43. A two-way communications transmission service that uses telephone or two-way cable is ______.
    (a) teletext
    (b) viewdata/videotex *
    (c) alphamosaic graphics
    (d) closed captioning
    (e) full definition video

44. The graphics technique used with videotex that produces displays that may seem boxy is ______.
    (a) full definition video
    (b) closed captioning
    (c) alphageometric graphics
    (d) alphamosaic graphics *
    (e) vertical blanking intervals

45. The Prestal videotex system of Great Britian uses ______ graphics.
(a) closed captioning
(b) full definition video
(c) alphamosaic *
(d) vertical blanking interval
(e) alpahgeometric

46. ______ graphics are generated by following instructions to create common shapes and locate them on the screen.
(a) Alphageometric *
(b) Closed captioning
(c) Vertical blanking interval
(d) Alphamosaic
(e) Full definition video

47. With ______ images will be prepared at the videotex computer and transmitted to the user.
(a) alphamosaic graphics
(b) full definition video *
(c) closed captioning
(d) alphageometric graphics
(e) vertical blanking intervals

Case Study 13-3: Viewtron

48. The videotex service offered by Knight-Ridder Newspapers is ______.
(a) Antiope
(b) Prestel
(c) Telidon
(d) Sceptre
(e) Viewtron *

49. The first viedotext service was ______ offered by the British Post Office.
(a) Viewtron
(b) Telidon
(c) Sceptre
(d) Prestel *
(e) Antiope

50. The system of viedotex that the French developed to replace the telephone book was ______.
(a) Telidon
(b) Viewtron
(c) Antiope *
(d) Prestel
(e) Sceptre

51. The Canadian videotex system developed to provide farm information is ______.
(a) Sceptre
(b) Telidon *
(c) Viewtron
(d) Antiope
(e) Prestel

52. The first commercially available videotex system offered in the U.S. was ______.
(a) Viewtron *
(b) Antiope
(c) Prestel
(d) Telidon
(e) Sceptre

On-Line Databases

53. The information providers that offer some of the same services as videotex, but especially for people with computer expertise are called ______.
(a) viewdata
(b) videotex
(c) closed captioning
(d) on-line databases *
(e) teletext

54. CompuServe is an example of a ______ service.
(a) teletext
(b) viewdata
(c) closed captioning
(d) videotex
(e) on-line database *

Social Impact

55. A major problem with the widespread use of computing is that of ______.
(a) education
(b) cost
(c) privacy *
(d) telecommuting
(e) communications

True/False

Introduction

1. *Time* magazine's "Man of the Year" in 1982 was Steve Jobs who was given the honor for creating the personal computer. (F p334)

A brief History of Personal Computing

2. The key development that made the personal computer possible was the invention of the microchip at Intel Corporation in 1971. (T p334)

3. The first model of the Apple computer built by Steve Jobs and Steve Wozniak became the machine practically equated with the term "personal computer." (F p335)

4. Radio Shack's first widely marketed personal computer was the Tandy-1. (F p336)

5. IBM entered the personal computer marketplace relatively late. (T p336)

6. Apple is the world's leading manufacturer of personal computers. (F p336)

Case Study 13-1: The IBM Personal Computer

7. The IBM Personal Computer was introduced in 1981 and rapidly became the world's leading personal computer. (T p336)

8. One key to the IBM PC hardware design is modularity--separate pieces may be combined in various ways to tailor a system to the needs of individual users. (T p337)

9. The system unit of the IBM PC contains the system board, the user memory, the permanent memory, and several expansion slots, among other items. (T p337)

10. The expansion unit provides a large capacity fixed disk drive. (T p338)

11. In the quantity 10Mb, the letters "Mb" mean megabits. (F p338)

12. The user memory in the IBM PC is made of ROM. (F p337)

13. On the IBM PC, user programs and portions of the operating system are executed in RAM. (T p338)

14. One advantage of floppy disks on the IBM PC over fixed disks is that they can hold much more information. (F p338)

15. A 10Mb fixed disk drive can hold the equivalent of over 5000 pages of double-spaced typewritten text. (T p338)

16. The keyboard of the IBM PC has both tactile and audio feedback. (T p338)

17. One problem with the IBM PC is that fast typists can easily "get ahead" of the machine's ability to digest incoming characters, and keystrokes are lost. (F p338)

18. One possible improvement in the IBM PC would be to add a "type-ahead" buffer. (F p338)

19. The monochrome display on the IBM PC is preferred over the color display by users who work mostly with text. (T p339)

20. The IBM graphics printer used with the IBM PC uses APA graphics; this means that it can set any dot on the page on or off. (T p339)

21. A printer with bidirectional printing can move the paper either forward or backward as it prints. (F p339)

22. The vast majority of software available for the IBM PC is supplied by IBM. (F p340)

Case Study 13-2: The Apple Macintosh

23. When the Macintosh was announced, its key technological advances were in sheer computing power and user-friendliness. (T p340)

24. The Macintosh uses the MC68000 8-bit microprocessor. (F p340)

25. With the Macintosh, most of the dialogue with the computer occurs with the use of a touch-sensitive screen. (F p341)

26. The use of the mouse, pointing, and icon graphics was pioneered by Apple and first offered with its Macintosh personal computers. (F p341)

27. The Macintosh screen is designed to resemble a businessperson's desktop. (T p341)

28. The Macintosh screen can display a control panel that allows the user to make adjustments in the background pattern, the sensitivity of the mouse, and so on; this makes the user interface even more friendly. (T p341)

29. The MacPaint software package turns the Macintosh into an artist's sketchpad. (T p344)

30. One problem with the Macintosh is that it is difficult to integrate both text and graphics into a single document. (F p344)

31. The comparatively large screen on the Macintosh contributes to the readability of displayed information. (F p344)

32. The Macintosh uses sturdy disks rather than floppy disks. (T p344)

33. One advantage of the IBM PC over the Macintosh is the PC's small "footprint." (F p345)

34. The disk surface on a sturdy disk is inaccessible until the disk is inserted into the drive, so dust and foreign matter are kept out. (T p345)

35. The Macintosh's companion printer, the Imagewriter, can print out everything that can be put on a Macintosh's screen. (T p347)

36. The Macintosh comes equipped with two RS-232/RS-422 ports that make it convenient to attach peripheral devices. (T p348)

37. RS-422 ports operate at 100 million bits per second, making it easy to hook a Macintosh into a local area network. (F p348)

38. The Apple Imagewriter printer used with the Macintosh actually has greater resolution than the screen. (T p348)

39. The Macintosh user can actually create customized type faces. (T p348)

40. The Macintosh provides an optional, low-cost sound generator that can produce four "voices" of music. (F p349)

Portable Computers

41. Most personal computers are portable. (F p349)

Choosing a Personal Computer

42. Most personal computers are sold through the mail. (F p349)

43. When searching for a personal computer it is wise to begin by writing down your needs. (T p350)

Personal Computer Data Communications

44. Modems enable personal computers to communicate over telephone lines. (T p351)

45. The RS-232 interface allows a personal computer to pass one record at a time to a device such as a modem, or to receive one record at a time from a device. (F p351)

46. In addition to special hardware, a data communications software package is normally needed

to enable personal computer data communications. (T p351)

Applications

47. Authors commonly use personal computers as word processors. (T p352)

48. Home computers facilitate working at home instead of traveling to an office; this is called telecommuting. (T p352)

Personal Computing Software

49. Today, computer hardware vendors produce most of the software used with their machines. (F p353)

50. Electronic spreadsheet software packages are used primarily in financial planning applications. (T p353)

51. One huge yet essentially untapped marketplace is expert systems applications for personal computer users. (T p353)

Education

52. One of the early uses of computers as educational tools was the replacement of slide rules with electronic calculators. (T p353)

53. Electronic calculators met with immediate acceptance in the classroom because they greatly speeded the educational process. (F p353)

54. Some educators are concerned that computers actually impede the educational process for students who do not feel comfortable working with the machines. (T p356)

55. One exciting use for computers in the classroom is simulation in which the computer might, for example, be used to reenact moments in history or even simulate space flights. (T p357)

56. Personal computers and laser videodisk players are combined to create interactive teaching machines. (T p357)

The Information Providers

57. In the future, our "information handling" problem is likely to change from obtaining useful information to sifting through mountains of it to find what we want. (T p357)

58. The failure of cable television has proven that people are unwilling to pay for various kinds of information brought to their homes and businesses by the communications media. (F p357)

59. The three main approaches offered by the companies called information providers are teletext, viewdata/videotex, and on-line databases. (T p357)

Teletext

60. Videotex repeatedly broadcasts a certain common base of information to its subscribers. (F p357)

61. The service providing teletext receives no communication back from the subscriber. (T p 357)

62. Teletext is transmitted in the black bands between the frames of a TV picture. (T p357)

63. Teletext is transmitted in vertical blanking intervals. (T p357)

64. Teletext systems can transmit a selection of about 100 pages or frames of text. (T p357)

65. The TV picture being transmitted loses some of its cripness when transmitted simultaneously with teletext. (F p357)

66. It is believed that teletext transmission over full cable TV channels would be an exciting and commercially viable product. (F p358)

67. Perhaps 5000 to 10,000 frames of teletext could be transmitted over a full cable TV channel. (T p358)

68. The first widespread use of videotex was for the closed captioning of TV programs for deaf viewers. (F p358)

69. One advantage of teletext is the broad selection of information available. (F p358)

Viewdata/Videotex

70. Videotex is called viewdata in Europe. (T p358)

71. Videotex is a much more powerful information service than teletext. (T p358)

72. Teletext is a two-way information service. (F p358)

73. Videotex communication is two-way with information sent over the telephone or a two-way cable TV channel. (T p358)

74. Today, most videotex communication is transmitted over two-way cable TV channels. (F p358)

75. With videotex, only the information specifically requested is transmitted. (T p358)

76. With videotex, the size of the database is limited by the bandwidth of the communications medium. (F p358)

77. With alphamosaic graphics, videotex images are generated as blocks of dot-matrix characters or graphics shapes. (T p358)

78. One problem with alphamosaic graphics is that the images may appear boxy. (T p358)

79. Great Britain's pioneering Prestel system uses alphageometric graphics. (F p358)

80. Alphagoemetric graphics are generated by following instructions to create certain common shapes and to locate them on the screen. (T p358)

81. Alphageometric graphics are generally of higher quality than alphamosaic graphics. (T p 358)

82. Alphageometric graphics are generally of higher quality than full-definition video images. (F p358)

83. Full definition video images require much greater bandwidths for transmission than alphageometric videotex images. (T p358)

84. Videotex is intended primarily for distribution to computer-literate people. (F p358)

85. Some telephone companies charge considerably higher rates for data transmission than for voice transmission. (T p358)

86. Today's relatively slow telephone transmission speeds hinder the effectiveness of videotex; this situation will improve with the use of two-way cable TV. (T p358)

87. One problem hindering the success of videotex is that it is used with TV sets that are already tied up with regular TV programming, cable TV, VCR tape players, videogames, personal computers, and the like. (T p358)

88. Much of what is availbale on today's videotex services can be obtained free from other sources, although perhaps not in as timely a fashion. (T p358)

Case Study 13-3: Viewtron

89. The Viewtron videotex service was developed in Great Britain. (F p359)

A Brief History

90. The first videotex service was Prestel, begun in 1979 by the British Post Office. (T p360)

91. The French Antiope videotex service was originally intended as a means of eliminating the expense of publishing and distributing telephone directories. (T p360)

92. The French expect to have a videotex terminal in the home of every person with a telephone by the early 1990s. (T p360)

93. The Australian government developed the Telidon system, which has been used mostly to provide agricultural information to farmers. (F p360)

94. The Viewtron system was first tested in 1983 with subscribers throughout South Florida. (F p360)

95. The first commercially available videotex system in the United States was Viewtron. (T p360)

How to Use Viewtron

96. Subscribers access Viewtron through Septre terminals provided by AT&T. (T p360)

97. One popular feature of Viewtron is that you can select up to 16 pages that you look at most often and put them in your personal magazine. (T p361)

98. Viewtron allows the subscriber to search an on-line encyclopedia. (T p362)

On-Line Databases

99. The on-line databases are offered by information providers to people with personal computers. (T p364)

100. On-line database services tend to be more user-friendly than videotex services. (F p364)

101. The premier on-line database services include CompServ, The Source, and the Dow Jones News/Retrieval Service. (T p364)

Fill-in-the-Blanks

1. **Personal** computers are devices that are more powerful than the huge million-dollar computers of 20 years age, and cost about as much as a TV set.

What Is a Personal Computer?

2. A **personal computer** is a compete computer system that is available for the "personal" use of one or more people at work or at home.

A Brief History Personal Computing

3. Steve Jobs and Steve Wozniak started the computer company known as **Apple**.

4. The first computer club in the United States was the **Home Brew** Computer Club.

5. The first computer marketed by the Tandy Corporation and sold through its network of Radio Shack retail stores was the **TRS-80**.

Case Study 13-1: The IBM Personal Computer

6. In 1981, IBM introduced the **IBM Personal Computer** and it rapidly became the world's most popular personal computer.

7. The hardware unit of the IBM-PC that houses the microprocessor, the RAM, and the ROM is the **system** unit.

8. The IBM graphics printer used with the IBM-PC, uses **APA (all-points-addressable)** graphics, which allows it to set any dot on the page on or off.

Case Study 13-2: The Apple Macintosh

9. In 1984 Apple introduced the **Macintosh** personal computer which is both powerful and extremely user-friendly.

10. With the Macintosh personal computer, most of the dialogue with the computer occurs with the use of a **mouse**, one of the most user-friendly input devices.

11. The Macintosh computer uses symbols called **icons** that describes the operation to be

performed.

12. The pushing of the button on the mouse of the Macintosh is called **clicking** .

13. The Macintosh computer uses a series of **pull-down** menus that display the choices available to the user.

14. The new disks used on the Macintosh are called **sturdy** disks.

Portable Computers

15. Most personal computers are not **portable** ; they can't easily be taken wherever a person goes.

Choosing a Personal Computer

16. **Computer** stores generally sell computer books, magazines, supplies and software packages; they provide repair service and conduct classes on how to use computers.

Personal Computer Data Communications

17. A personal computer must have both a modem and an **RS-232 interface** in order to communicate over telephone lines.

Applications

18. **Word processing** refers to the typing, editing, and printing of text on a computer.

19. The personal computer software packages that help people organize household budgets, maintain Christmas card lists, and store inventories of household valuables are **database management** packages.

Personal Computing Software

20. Expert advice in such topics as taxes, legal problems, and first aid are available with software called **expert systems** .

Education

21. With **computer assisted instruction (CAI)** , learning is aided by computers.

The Information Providers

22. Commercial enterprises called **information providers** distribute information electronically; these include CompuServe and The Source.

Teletext

23. **Teletext** is a one-way transmission service, it repeatedly broadcasts a certain common base of information.

24. Teletext information is transmitted in the black bands often seen on the TV screen; these bands are called **vertical blanking** intervals.

25. The first wide spread use of teletext was for the **closed captioning** of TV programs for deaf viewers.

Viewdata/Videotex

26. **Videotex** , called viewdata in Eurrope, is a much more powerful information service than teletext.

27. **Videotex** is a two-way information service sent over the telephone or a cable TV channel.

28. Videotex graphics produced with **alphamosaic** graphics may tend to appear boxy.

29. The videotex graphics produced with **alphageometric** graphics are generated by following instructions to create certain common shapes and to locate them on the screen.

30. Future videotex graphics will be **full definition video images** that will be prepared at the videotex computer and transmitted to the user; these will be extremely sharp and realistic.

Case Study 13-3: Viewtron

31. The first videotex service was **Prestel** , begun in 1979 by the British Post Office.

32. The French developed the **Antiope** service, that was intended to replace the publishing of phone directories.

33. **Telidon** is the system developed by the Canadian government to provide agricultural information.

34. The first commercially available videotex service in the United States was **Viewtron** .

How to Use Viewtron

35. Subscribers access Viewrton through **Sceptre** terminals provided by AT&T.

On-Line Databases

36. The information providers that offer services to people with personal computers are called **on-line databases** .

37. Three of the best known on-line databases are **CompuServe** , The Source, and the Dow Jones News/Retrieval Service.

Social Impact

38. PCs enable us to **telecommute** , that is to work at home by using computers and communications thus reducing time-consuming and costly commuting.

Chapter 14
Office Automation

Multiple Choice

1. ______ refers to the use of computing and communications technologies to process office information electronically.
   (a) FAX
   (b) PABX
   (c) OA *
   (d) OCR
   (e) CBX

2. The introduction of word processing started the field of ______.
   (a) CBX
   (b) FAX
   (c) OCR
   (d) PABX
   (e) OA *

3. ______ integrates the processing of numbers, text, graphics, voice, and video images.
   (a) OCR
   (b) PABX
   (c) FAX
   (d) OA *
   (e) CBX

4. One of the primary goals of ______ is the improvement of the overall communications flow in an organization.
   (a) OA *
   (b) FAX
   (c) PABX
   (d) CBX
   (e) OCR

Goals of Office Automation Systems

5. Some people are hoping to develop ______ where an organization's information is entirely produced and stored electronically.
   (a) stand-alone word processors
   (b) paperless offices *

(c) teleconferencing
(d) centralized control
(e) gateways

6. One goal of office automation is to enable each worker to work from one location called a computerized ______.
(a) word processor
(b) gateway
(c) facsimile
(d) optical character recognition
(e) workstation *

Characteristics of Offices and OA Systems

7. Workstations must provide ______ interfaces.
(a) word processing
(b) teleconferencing
(c) user-friendly *
(d) data communication
(e) modular

8. OA systems should be ______, consisting of separate pieces that can be combined to solve each user's particular problems.
(a) integrated
(b) centrally controlled
(c) distributed controlled
(d) modular *
(e) stand-alone

9. ______ is the term used to describe OA systems in which all pieces must work together.
(a) Integration *
(b) Stand-alone
(c) Modular
(d) Centrally controlled
(e) Distributed control

10. With the ______ feature, a user can cancel a current activity and go back to the previous activity.
(a) Redo
(b) Undo *
(c) Todo
(d) Dodo
(e) Cancel

11. Many OA stations use ______ to keep track of several tasks at once.

(a) menu bars
(b) encryption
(c) insertion points
(d) global replace
(e) windowing *

Word Processing

12. A ______ is a computer-based system that creates and manipulates text.
(a) multilingual dictionary
(b) modular system
(c) word processor *
(d) electronic calendar
(e) facsimile

13. A ______ word processor is a complete single-user system.
(a) shared-resource
(b) facsimile
(c) shared-logic
(d) stand-alone *
(e) gateway

14. With a ______ system, several terminals use the same processor and peripheral devices.
(a) shared-logic *
(b) facsimile
(c) gateway
(d) shared-resource
(e) stand-alone

15. If the central processor fails in a ______ system, the entire system fails.
(a) gateway
(b) shared-logic *
(c) stand-alone
(d) shared-resource
(e) facsimile

16. In a ______ system, each workstation has its own processor.
(a) shared-logic
(b) facsimile
(c) shared-resource *
(d) stand-alone
(e) gateway

17. With a ______ system, if one workstation fails the entire system does not fail.
(a) facsimile

(b) gateway
(c) stand-alone
(d) shared-resource *
(e) shared-logic

18. An office typewriter can become an effective input device when used with a(n) ______ page reader.
(a) facsimile
(b) CBX
(c) PABX
(d) FAX
(e) OCR *

Capabilities of Word Processing Systems

19. The feature of a word processor that moves text typed over a margin down to the next line is ______.
(a) scrolling
(b) word wraparound *
(c) cut and paste
(d) right-justification
(e) automatic hyphenation

20. The feature of a word processor that allows a user to view documents wider than the screen is the ______.
(a) horizontal scroll *
(b) vertical scroll
(c) right-justification
(d) automatic hyphenation
(e) word wraparound

21. ______ feature of a word processor enables a used to move forward or backward in a document.
(a) Right-justification
(b) Automatic hyphenation
(c) Vertical scrolling *
(d) Word wraparound
(e) Horizontal scrolling

22. The feature of a word processor that allows a user to remove text from one page and insert it on another page is called ______.
(a) global replace
(b) undo
(c) list processing
(d) cut and paste *

(e) prestored phrases

23. A user can change all occurences of a word or phrase with the _____ feature of a word processor.
(a) cut and paste
(b) redo
(c) undo
(d) local replace
(e) global replace *

24. With the _____ feature of a word processor, a user can have both margins on each side of a page even.
(a) horizontal scrolling
(b) right-justification *
(c) automatic hyphenation
(d) word wraparound
(e) cut and paste

25. With the _____ feature of a word processor, frequently used texts are stored in the computer.
(a) automatic page numbering
(b) cut and paste
(c) prestored phrases *
(d) automatic hyphenation
(e) multilingual dictionaries

26. Word processors with the _____ feature can generate mass mailings.
(a) electronic dictionaries
(b) prestored phrases
(c) cut and paste
(d) list processing *
(e) headings

27. Spelling can be checked by word processors that have a(n) _____ feature.
(a) electronic dictionary *
(b) automatic hyphenation
(c) list processing
(d) automatic page numbering
(e) archival storage

28. Word processors with electronic dictionaries can perform _____.
(a) automatic page numbering
(b) archival storage
(c) list processing
(d) automatic hyphenation *
(e) global replace

29. A word processor with ______ can convert text from one language to another language.
    (a) electronic dictionaries
    (b) archival storage
    (c) automatic hyphenation
    (d) list processing
    (e) multilingual dictionaries *

Case Study 14-1: Word Processing with Apple's Macintosh

30. The display of the various icons on the Macintosh is often called a ______.
    (a) insertion point
    (b) desktop *
    (c) ruler
    (d) scrapbook
    (e) I-beam

31. Pressing the button on the mouse of the Macintosh is called ______.
    (a) pulling down
    (b) dragging
    (c) clicking *
    (d) inserting
    (e) tickling

32. ______ occurs on the Macintosh while moving the mouse and holding down its button.
    (a) Clicking
    (b) Opening
    (c) Inserting
    (d) Dragging *
    (e) Tickling

33. The blinking bar on the Macintosh indicates the ______.
    (a) insertion point *
    (b) menu
    (c) clipboard
    (d) ruler
    (e) desktop

34. When the ______ key is hit on the Macintosh the insertion point moves down to the next line.
    (a) Redo
    (b) Shift
    (c) Backspace
    (d) Tab
    (e) Return *

35. With the Macintosh, the Undo feature can be undone by selecting ______.
    (a) Edit
    (b) Redo *
    (c) Save
    (d) Eject
    (e) Quit

36. A session on the Macintosh is ended by selecting ______.
    (a) Finish
    (b) Eject
    (c) Quit *
    (d) Exit
    (e) Return

37. To remove the disk from the disk drive on the Macintosh the user selects ______.
    (a) Quit
    (b) Edit
    (c) Return
    (d) Eject *
    (e) Font

38. To continue in MacWrite but wrap up a current document the user selects ______.
    (a) Eject
    (b) Quit
    (c) Style
    (d) Edit
    (e) Close *

39. Text removed from a document in MacWrite is placed in an area called ______
    (a) File
    (b) Clipboard *
    (c) Scrapbook
    (d) Save
    (e) Edit

40. To change a single piece of text in word processing is called a ______.
    (a) local replace *
    (b) edit
    (c) global replace
    (d) cut
    (e) draft

41. Frequently used text selections can be stored in MacWrite in the ______.
    (a) scroll bar
    (b) Clipboard

(c) Scrapbook *
(d) Draft
(e) Folder

Electronic Mail

42. _____ is correspondence that is sent via computer and communication systems.
(a) Broadcast
(b) Data compression
(c) Tickler message
(d) Electronic mail *
(e) Encryption

43. The telegram was an early form of _____.
(a) facsimile
(b) tickler message
(c) broadcast
(d) teleconferencing
(e) electronic mail *

44. Electronic messages that cannot be delivered immediately are stored in _____.
(a) clipboard
(b) electronic mailboxes *
(c) voiceprints
(d) electronic blackboards
(e) scrapbook

45. Electronic messages are sent to a group of people called a _____.
(a) teleconference
(b) committee
(c) mail group
(d) distribution list *
(e) broad group

46. The coding of messages is called _____.
(a) encryption *
(b) facsimile
(c) decryption
(d) gateways
(e) broadcasting

47. _____ refers to the uncoding of messages.
(a) Broadcasting
(b) Gateways
(c) Decryption *

(d) Facsimile
(e) Encryption

Facsimile

48. ______ devices are essentially long-distance copiers.
   (a) PABX
   (b) FAX *
   (c) OA
   (d) OCR
   (e) CBX

49. With ______ a hardcopy duplicate of an original document can be transmitted long distances.
   (a) electronic mailboxes
   (b) telephone tag
   (c) teleconferencing
   (d) electronic scheduling systems
   (e) facsimile *

Voice Mail

50. With ______, messages are sent as human speech.
   (a) voice mail *
   (b) electronic mail
   (c) facsimile
   (d) telephone tag
   (e) call detail reporting

51. The term associated with people constantly calling one another but never connecting is ______
   (a) voice mail
   (b) conference calls
   (c) telephone tag *
   (d) call notification
   (e) station hunting

52. A reminder that someone can leave for themself on a voice mail system is called a ______.
   (a) automatic call back
   (b) tickler message *
   (c) voiceprint
   (d) gateway
   (e) call notification

Electronic Calendar

53. A ______ would contain confirmed meetings, tentative meetings, and open time slots for each person in a organization.
    (a) electronic mailbox
    (b) distribution list
    (c) electronic in-basket
    (d) electronic calendar *
    (e) gateway

54. Using a ______ system, meetings can be scheduled well in advance along with all the equipment that might be needed.
    (a) gateway
    (b) distribution list
    (c) electronic mailbox
    (d) electronic in-basket
    (e) electronic calendar *

Teleconferencing

55. ______ incorporates audio, video, and computer conference techniques.
    (a) Station hunting
    (b) Teleconferencing *
    (c) Electronic scheduling
    (d) Electronic blackboard
    (e) Electronic mail

56. With ______, the picture is regenerated approximately 30 times per second.
    (a) freeze-frame videoconferencing
    (b) facsimile
    (c) continuous motion videoconferencing *
    (d) electronic blackboards
    (e) electronic scheduling

57. Using ______ videoconferencing, participants see what appears to be a slide presentation.
    (a) freeze-frame *
    (b) continuous motion
    (c) blackboard
    (d) desk-to-desk
    (e) voice

Electronic Blackboards

58. Using ______ the changing positions of the chalk are transmitted via telephone.

(a) freeze-frame videoconferencing
(b) desk-to-desk videoconferencing
(c) electronic blackboards *
(d) continuous motion videoconferencing
(e) chalk conferencing

OA Communications: The Computerized Branch Exchange

59. ______ are devices that carry both voice and data.
(a) Electronic blackboards
(b) Computerized branch exchanges *
(c) Facsimile
(d) Data compression
(e) Tickler messages

60. The feature of a CBX that locates the most economical way to make a call is ______.
(a) station hunting
(b) call detail reporting
(c) callback queueing
(d) least cost routing *
(e) automatic call back

61. The ______ feature of a CBX will call back when a line is free to use.
(a) automatic call back
(b) call forwarding
(c) least cost routing
(d) station hunting
(e) callback queueing *

62. The CBX feature that lets a person know there is another call waiting is ______.
(a) call forwarding
(b) call notification *
(c) call back queueing
(d) automatic callback
(e) station hunting

63. The CBX feature that helps management keep track of unauthorized use of the phones is ______.
(a) station hunting
(b) call back queueing
(c) call detail reporting *
(d) call notification
(e) automatic callback

64. Different types of networks can be interconnected by using ______.

(a) gateways *
(b) conference calls
(c) LANs
(d) PABXs
(e) station hunting

Computerized Typesetting

65. In computerized typesetting, special ______ must be inserted into the text that tell the equipment the size and style of type and spacing required.
(a) gateways
(b) access codes
(c) encryption codes
(d) control codes *
(e) decryption codes

66. Computerized typesetting systems are also called ______ systems.
(a) data compression
(b) photocomposition *
(c) shared-logic
(d) shared resource
(e) word processing

Archival Storage

67. Office automation systems can provide ______ for older and less active documents.
(a) aged storage
(b) encryption
(c) archival storage *
(d) data compression
(e) control storage

68. Computer output microfilm is a popular medium for ______.
(a) word processor
(b) shared-logic system
(c) shared resource system
(d) stand-alone system
(e) archival storage *

Obstacles to the Growth of Office Automation

69. Information flow can be slowed down because executives ______.
(a) do not use the system *

(b) fear they may lose their jobs
(c) are reluctant to spend money for OA systems
(d) feel that the equipment will become out-of-date
(e) that OA systems will hamper smooth running offices

Trends in Office Automation

70. With ______, pictures of objects will be stored on magnetic media rather than film.
(a) laser technology
(b) color CRTs
(c) omnifont readers
(d) digital photography *
(e) computer graphics

True/False

Introduction

1. More than half the work force in the United States consists of white-collar workers, people whose primary job is the manipulation of information. (T p370)

2. White-collar work is becoming increasingly labor-intensive. (F p370)

3. Office automation began in the mid-1960s with the introduction of database management systems. (F p370)

Goals of Office Automation Systems

4. One goal of office automation systems is allowing a person's work to be handled from a single location. (T p371)

5. One goal of office automation systems that is near being realized is the "paperless office." (F p371)

Characteristics of Offices and OA Systems

6. One key to the success of OA is integration, that is, all the pieces of OA equipment must work together. (T p372)

7. Most OA systems today offer a menu approach in which the alternatives available to the user at any point are displayed on a screen. (T p372)

8. A recent innovation in office automation is the UNDO key that allows a user to cancel the effects of the current activity and go back to the previous activity. (T p372)

9. THe HELP key is commonly known as the "second-chance" key. (F p372)

10. Windowing is a common technique used on office automation workstations to help a user keep track of several tasks at once. (T p372)

Word Processing

11. Xerox Corportaion pioneered the integration of the mouse, windowing, and icon graphics with the development of the Star Information System terminals. (T p372)

12. Richard Nixon was the first president to prepare his memoirs on a word processor. (F p373)

13. The leading technology in office automation is, and always has been, electronic mail. (F p374)

14. A stand-alone word processor is a complete single-user system. (T p374)

15. Each word processing station in a shared-logic system has its own processor, but the peripheral devices are shared. (F p374)

16. Each word processing station in a shared-resource system shares the central processor, storage, and peripheral devices. (F p374)

17. In a shared-resource word processing system, a failure in one work station does not affect the others. (T p374)

18. The common typewriter becomes an effective input device when used with an OMR reader. (F p374)

Capabilities of Word Processing Systems

19. With the word wraparound feature in a word processing system, a word typed over the right margin is automatically split and hyphenated. (F p375)

20. A word processor with horizontal scrolling enables the user to create documents wider than the screen. (T p375)

21. With vertical scrolling, a word processor can move forward or backward in a document. (T p375)

22. A word processor with cut-and-paste capability allows the user to form documents from individual pieces and to insert text and graphics at any point in a document. (T p375)

23. The local replace capability allows the user to change all occurrences of a word or phrase in a document to a new word or phrase. (F p375)

24. Some word processors offer a capability called list processing that allows the user to generate mass mailings. (T p375)

25. Word processors with electronic dictionaries automatically locate and correct misspelled words. (F p375)

26. Word processors with electronic dictionaries may perform automatic hyphenation. (T p375)

27. The proper way to hyphenate the word "project" is clearly " pro-ject." (F p375)

28. The IBM Displaywriter has a built-in electronic dictionary that is used to verify the spelling of words in a document at over 1000 word per minute. (T p376)

29. Most word processors offer multilingual dictionaries that accurately translate text from one language to another. (F p376)

Case Study 14-1: Word Processing with Apple's Macintosh

30. Pressing the button on the mouse once and releasing it quickly is called clicking. (T p376)

31. On the Macintosh, clicking selects the item being pointed to. (T p376)

32. Moving the mouse whlie the button is held down is called dragging. (T p376)

33. With the Macintosh, a word typed across the right margin is automatically moved to the next line; this is called word wraparound. (T p378)

34. With MacWrite, at any time you can undo what you have done since the last click of the mouse by selecting Redo from the Edit menu. (F p379)

35. With MacWrite, the Undo itself can be undone by reselecting Undo from the Edit menu. (F p379)

36. An important reason for the speed of MacWrite is the great speed of the Motorola 68000 microprocessor chip at the heart of the Macintosh. (T p379)

37. MacWrite automatically hyphenates words when it performs full justification. (F p379)

38. In MacWrite, text is ordinarily selected by dragging the pointer across it. (T p380)

39. MacWrite highlights selected text by underlining it. (F p380)

40. In MacWrite, plaintext, bold, italic, underline, outline, and shadow are examples of text fonts. (F p380)

41. In MacWrite, text styles may be combined to create interesting effects like bold italic. (T p380)

42. Each new MacWrite document is ordinarily preset to 12-Point Plain Text type. (T p380)

43. In MacWrite, New York, Geneva, Toronto, Monaco, Chicago, Venice, London, Athens, and San Francisco are examples of type styles. (F p380)

44. Chicago is the typeface used by the Macintosh itself in its various screen presentations. (T p380)

45. The Geneva typeface is ordinarily preset in each new MacWrite document. (F p380)

46. In MacWrite, deletions may be made by selecting the text and pressing the backspace key once. (T p380)

47. In MacWrite, selecting the full-justification box on the ruler causes the text to have even left and right margins. (T p381)

48. In MacWrite, text that is Cut is automatically placed in the Scrapbook, a temporary holding area. (F p282)

49. In MacWrite, by choosing Paste from the File menu, the text from the Clipboard is moved to the insertion point. (F p382)

50. In MacWrite, to copy a block of text from one point in the document to another, select the text to be copied, choose Copy from the Edit menu, select the insertion point, and choose Paste from the Edit menu. (T p382)

51. In MacWrite, clicking Change All causes a stern warning to appear because this operation is not undoable. (T p382)

52. MacWrite provides a feature called the Clipboard in which many frequently used text selections may be stored. (F p384)

53. In MacWrite the Scrapbook is located via the Apple menu. (T p384)

Electronic Mail

54. The Western Union Mailgram was the earliest form of electronic mail. (F p384)

55. Another name for an electronic mailbox is an electronic in basket. (T p384)

56. Electronic mail may be sent to a select group called a distribution list. (T p384)

57. An electronic mail system broadcasts a message to all the people in a distribution list. (T p384)

58. Facsimile transmission is a form of electronic mail. (T p385)

59. FAX devices are essentially long-distance copiers. (T p385)

60. Data compression is a technique used to speed facsimile transmission. (T p385)

Voice Mail

61. With voice mail, human voice is converted to text and then transmitted. (F p385)

62. Voice mail is delivered as speech, exactly as the caller spoke it. (T p385)

63. Most of today's voice mail systems store voiceprints of their users; this enables them to verify the author of a voice message. (F p385)

64. One serious problem with voice mail systems is that they cannot relate the author's mood. (F p385)

65. "Telephone tag" occurs when two people are constantly calling and returning calls but can't seem to connect. (T p386)

66. If your phone rings and it's you calling, you have probably sent yourself a tickler message. (T p386)

67. One potential problem with voice mail messages is that they might be altered by impersonators. (T p386)

Electronic Calendar

68. Electronic calendars and electronic diaries are conveniences of the electronic office that could violate an individual's privacy. (T p387)

Teleconferencing

69. With continuous motion videoconferencing, the picture is regenerated approximately 300 times per second. (F p387)

70. With freeze-frame videoconferencing, the conference participants see what appears to be a slide show presentation with a new slide displayed every several seconds. (T p388)

71. Freeze-frame videoconferencing requires much greater bandwidth than continuous motion videoconferencing. (F p388)

72. Freeze-frame videoconferencing has the advantage that it can be transmitted over conventional telephone lines. (T p388)

73. Desk-to-desk teleconferencing has become commonplace in today's automated offices. (F p388)

Electronic Blackboard

74. The electronic blackboard is pressure sensitive so it knows where the chalk is at any moment. (F p388)

75. With the electronic blackboard,, information is transmitted by telephone to remote locations where it is displayed on TV monitors. (T p388)

OA Communications: The Computerized Branch Exchange

76. Wangnet is the most widely used baseband network. (F p388)

77. The trend in local area networks is from baseband networks to broadband networks. (T p388)

78. Today's CBXs carry both voice and data. (T p388)

79. A synonym for PBX is CBX. (F p388)

80. With least cost routing, the CBX automatically locates the most economically available path for making a call. (T p388)

81. With callback queueing, a person finding a particular line busy is automatically called back when the line becomes available. (T p388)

82. With call notification, if the party being called is already talking on the phone, the CBX calls that number back automatically when the party hangs up. (F p388)

83. With call forwarding, the CBX can automatically forward calls for someone who has moved to another number. (T p389)

84. With station hunting, the CBX automatically routes the call through the least expensive available line. (F p389)

85. With the conference calls feature, the CBX automatically arranges teleconferences. (F p389)

86. With call detail reporting, management can keep track of all calls initiated from each extension. (T p389)

87. With automatic callback, the CBX rings a caller back when a busy party hangs up. (T p389)

88. Confirmation of the value of the CBX concept in office automation has come from Wang's investment in Rolm. (F p389)

89. Most of today's CBXs operate at speeds of one million bps. (F p389)

90. Most of today's CBXs are almost indistinguishable from LANs. (F p389)

91. Today's baseband LANs commonly operate at speeds of 10 billion bps or more. (F p389)

92. Today's broadband LANs generally have far greater capacity than today's baseband LANs. (T p389)

93. CBXs have centralized control because of their star topology. (T p389)

94. If the CBX fails, the entire communications system fails. (T p389)

95. Most LANs use centralized control. (F p389)

96. In most LANs, if one device fails the remainder of the network continues functioning. (T p389)

97. In the near term, it is clear that companies should use CBXs and avoid LANs. (F p389)

98. CBXs and LANs may be interconnected by devices called gateways. (T p389)

99. One disadvantage of hybrid networks combining LANs and CBXs is that the networks can operate only as fast as their slowest components. (T p389)

100. It seems clear that the broadband LAN will eventually become the premier office automation communications medium. (T p389)

Computerized Typesetting

101. The computerized typesetting process is not quite as simple as it sounds because many special control codes must be incorporated into the text. (T p389)

102. Computerized typesetting systems are also called photocomposition systems. (T p390)

103. One popular form of photocomposition system operates as follows: Character images are generated on a high-resolution CRT; the CRT image is then photographed to produce the film containing the typest material. (T p390)

Archival Storage

104. Office automation systems provide archival storage for older and less active documents. (T p390)

105. Computer output microfilm is popular for archival storage of information that will need to be read back into a computer. (F p390)

106. Computer tape is ordinarily used for archiving information that may eventually require futher computer processing. (T p390)

Obstacles to the Growth of Office Automation

107. One problem with investing in OA systems is that computer equipment in recent years has become obsolete quickly. (T p390)

108. The fact that many executives can't type has hindered the acceptance of office automation. (T p390)

Trends in Office Automation

109. One reason the CBX approach may at least temporarily win over LANs is that most existing buildings are already wired for telephone communications. (T p392)

Fill-in-the-Blanks

1. **Office automation (OA)** refers to the use of the computing and communications technologies to process information in offices electronically.

2. Computers that manipulate textual material such as words, phrases, sentences, and paragraphs are **word processing** systems.

Goals of Office Automation Systems

3. One of the goals of office automation is the **paperless office**, in which all documents would be produced and filed electronically eliminating paper copies.

4. Computerized **workstations** in an OA system would allow workers to transact all their business from one location.

Characteristics of Offices and OA Systems

5. One of keys to the success of OA is **integration**, that is, all the pieces of OA equipment must work together.

6. The **"HELP"** key on an OA station allows the user to ask for assistance.

7. The **"UNDO"** key on a OA station allows a user to cancel the effects of a current activity and go back to the previous activity.

8. Many OA stations use **windowing** to keep track of several tasks at once.

Word Processing

9. A **word processor** is a computer-based system that facilitates the creation, storage, editing, transmission, and display or printing of text.

10. A **stand-alone** word processor is a complete single-user system, containing a keyboard, a screen, a printer, and disk storage.

11. Each station in a **shared-logic** word processing system shares the central processor, storage, and peripheral devices.

12. In a **shared-resource** word processing system each workstation has its own processor, and the peripheral devices are shared.

Capabilities of Word Processing Systems

13. The feature of a word processor that automatically moves a word to the next line when it is types over a margin is **word wraparound**.

14. The **horizontal scrolling** feature of a word processor allows a user to view portions of a document wider than the screen.

15. The **vertical scrolling** feature of a word processor allows a user to move forward or backward in a document.

16. The use of the **cut-and-paste** feature of a word processor allows the user to take pieces from one document and insert them into another document.

17. The **global replace** feature of a word processor allows the user to change all occurrences of a word or phrase in a document to a new word or phrase.

18. Most word processors allow the user to specify **automatic page numbering** in which page numbers are automatically calculated and inserted on each page.

19. The feature of a word processing system that stores often used portions of text is called **prestored phrases** .

20. Some word processors offer a capability called **list processing** that helps the user to generate mass mailings.

21. With the **electronic dictionary** feature on a word processor spelling checks are made on each word in a document.

22. Some word processors provide **multilingual dictionaries** to help a user convert text from one language to another.

Case Study 14-1: Word Processing with Apple's Macintosh

23. The word processing software supplied by Apple for the Macintosh computer is called **MacWrite** .

24. When starting a MacWrite document, a menu bar and several icons appear on the screen, this is called the initial **desktop** .

25. The **clicking** of the button on the mouse of the Macintosh computer selects the item being pointed to.

26. To pull down a menu on the Macintosh computer, the user holds down the mouse button while moving the mouse in what is called **dragging** .

27. MacWrite uses a blinking bar to mark the **insertion point** , the location in a document where text may be placed.

28. The Undo feature in MacWrite can de undone by selection **Redo** from the Edit menu.

29. The **Clipboard** is a temporary holding area in MacWrite where pieces of text are placed when they have been Cut from a document.

30. The feature of MacWrite that stores many frequently used text selections is called the **Scrapbook** .

Electronic Mail

31. With **electronic mail** , correspondence is sent electronically rather than by the movement of paper.

32. An **electronic mailbox** is sometimes called an electronic in basket.

33. Electronic mail can be sent to a select group called a **distribution list** .

34. In an electronic mail system, the sender specifies who is to receive a message, and the system **broadcasts** the message only to those people.

35. **Encryption** of electronic mail messages prevents unauthorized users from reading user communications obtained by tapping a system.

36. When a message has been encrypted, only authorized users have access to the system's **decryption** facilities for decoding purposes.

Facsimile

37. **Facsimile (FAX)** devices are essentially long-distance copiers.

Voice Mail

38. **Voice mail** is a form of electronic mail in which messages are sent as human voice rather than streams of text.

39. Future voice mail systems will store **voiceprints** of their users that will enable the systems to verify user authorization.

40. On a voice mail system, users can send reminders to themselves; these are called **tickler** messages.

Electronic Calendar

41. Each person in an automated office may maintain an electronic **calendar** containing confirmed meetings, tentative meetings, and open time slots.

Teleconferencing

42. **Teleconferencing** incorporates audio, video, and computer conferencing techniques.

43. The form of teleconferencing that provides much of the atmosphere of personal contact is **videoconferencing** .

44. With **continuous motion** videoconferencing, the picture is regenerated approximately 30 times per second.

45. With **freeze-frame** videoconferencing, the conference participants see what appears to be a slide show presentation.

46. As communications companies increase transmission bandwidths and reduce costs, **desk-to-desk** teleconferencing may eventually become commonplace.

Electronic Blackboard

47. Electronic **blackboards** digitize and transmit the patterns being drawn by chalk.

OA Communications: The Computerized Branch Exchange

48. **Computerized branch exchanges (CBXs)** , sometimes called PABXs, are devices that carry both voice and data.

49. The feature of a CBX that automatically checks for the most economical phone lines is called **least cost routing** .

50. The **callback queueing** feature of a CBX automatically calls a person back when a line becomes available.

51. The **call forwarding** feature of a CBX will automatically transfer calls to a new number if a person has moved.

52. With the **station hunting** feature of a CBX, a call to a busy extension can be routed to a nearby extension.

53. The CBX feature that allows three or more parties to take part in the same call is called **conference calls** .

54. The CBX feature that allows management to to keep track of all calls initiated from each extension is called **call detail reporting** .

55. With **automatic callback** , the CBX rings a caller back when a busy party hangs up.

56. CBXs have the disadvantage of **centralized** control because of their star topology; if the CBX fails the entire communications system fails.

57. Local area networks (LANs), generally use **distributed** control; if one device fails, the remainder of the network continues functioning.

58. Fundamentally different kinds of communications networks may be interconnected by devices called **gateways** .

Computerized Typesetting

59. With computerized typesetting, special **control codes** must be inserted to tell the typesetting machine what size, style, and font of text to use.

60. Computerized typesetting systems are also called **photocomposition** systems.

Archival Storage

61. Office automation systems provide **archival** storage for older and less active documents.

Chapter 15
Electronic Funds Transfer, Security, Privacy, and Computer Crime

Multiple Choice

1. ______ refers to the movement of money by computers and data communications systems.
   (a) Automated clearing houses
   (b) Electronic funds transfer *
   (c) On-line teller terminals
   (d) Automated teller machines
   (e) National Commission on Electronic Funds Transfers

2. The automatic payment of loans is referred to as ______.
   (a) pay by phone
   (b) bank-at-home
   (c) point-of-sale
   (d) preauthorized computer-initiated payments *
   (e) credit authorization

3. A ______ society refers to having all financial transactions made electronically.
   (a) cashless/checkless *
   (b) debit card
   (c) bank-at-home
   (d) pay-by-phone
   (e) less cash/less check

Direct Deposit

4. One of the earliest forms of EFT was ______.
   (a) bank-at-home
   (b) point-of-sale
   (c) direct deposit *
   (d) on-line teller terminals
   (e) preauthorized computer-initiated payments

5. With ______ an employee's pay can be placed into a bank account without the need of a paycheck.

(a) on-line teller terminals
(b) preauthorized computer-initiated payments
(c) bank-at-home
(d) automated clearing houses
(e) direct deposit *

6. The computer centers that handle electronic fund transfers are called ______.
   (a) automated clearing houses *
   (b) automated teller machines
   (c) preauthorized computer-initiated payments
   (d) credit authorization terminals
   (e) Federal Reserve

7. The Federal Reserve Board controls ______ that handle electronic fund transfers between banks.
   (a) preauthorized computer-initiated payments
   (b) automated clearing houses *
   (c) credit authorization terminals
   (d) debit cards
   (e) automated teller machines

On-Line Teller Terminals

8. ______ are machines used at banks to enter information directly into the banks' computers.
   (a) Point-of-sale terminals
   (b) Credit authorization terminals
   (c) Automated clearing terminals
   (d) On-line teller terminals *
   (e) Pay-by-phone terminals

Automated Teller Machines

9. With ______, bank transactions can be handled without the need for human tellers.
   (a) point-of-sale terminals
   (b) credit authorization terminals
   (c) automated teller machines *
   (d) automated clearing houses
   (e) on-line teller terminals

10. To operate an ATM a bank customer must have an account with the bank, an identification card, and a ______.
   (a) access key
   (b) pass key
   (c) voiceprint

(d) credit card
(e) personal identification number *

11. Which of the following is not a disadvantage of ATMs? ______
(a) Bank employees are not endangered in a robbery *
(b) They are targets for vandalism
(c) They are costly to install
(d) Correcting errors is difficult
(e) They tend to malfunction

Box: Using ATMs

12. When selecting a ______ for use at an ATM, it is important not to pick something obvious.
(a) debit card
(b) pass key
(c) PIN *
(d) account number
(e) decrryption key

Credit Authorization Terminals

13. With the use of ______ terminals merchants can determine if a check or a customer's credit is alright.
(a) automated teller
(b) point-of-sale
(c) pay-by-phone
(d) credit authorization *
(e) on-line teller

14. ______ terminals are connected by phone to networks of computers that contain current credit information.
(a) On-line teller
(b) ATM
(c) Automated teller
(d) Point-of-sale
(e) Credit authorization *

The Debit Card and The Credit Card

15. When using a ______, a person is borrowing money.
(a) PIN
(b) credit card *
(c) access code

(d) debit card
(e) POS

16. With the use of a ______, funds are transfered immediately from one account to another over the EFT networks.
(a) POS
(b) credit card
(c) debit card *
(d) PIN
(e) ATM

Pay-by-Phone

17. Banks offer ______ for clients with touch-tone phone service.
(a) preauthorized computer-initiated payments
(b) bank-at-home
(c) credit authorization
(d) pay-by-phone *
(e) direct deposit

18. With the use of ______ people can authorize a bank to pay certain bills without having to write a check.
(a) pay-by-phone *
(b) direct deposit
(c) credit cards
(d) ATMs
(e) point-of-sale terminals

Point-of-Sale Terminals and Transactions

19. With the use of ______ terminals, stores can enter item and customer information directly into their computers to speed the check-out process.
(a) ATM
(b) on-line
(c) PIN
(d) credit authorization
(e) POS *

20. ______ terminals are used with UPC scanners to speed the check-out process in many stores.
(a) PIN
(b) POS *
(c) Debit card
(d) ATM

(e) Credit authorization

Privacy and Social Issues in EFT

EFT Legislation and Regulation

21. The group established by Congress to recommend legislation on how EFT systems should be monitored and controlled was the ______.
    (a) NCIC
    (b) EFT
    (c) NCEFT *
    (d) EBCDIC
    (e) ANSI

22. The recommendations of the NCEFT were implemented by the ______.
    (a) Privacy Act
    (b) Right to Financial Privacy Act
    (c) Freedom of Information Act
    (d) Electronic Funds Transfer Act *
    (e) Fair Credit Reporting Act

23. ______ implements the provisions of the Electronic Funds Transfer Act.
    (a) Regulation E *
    (b) NCEFT
    (c) Privacy Act
    (d) Fair Credit Reporting
    (e) Right to Financial Privacy Act

The Future of EFT

Computer Crime Techniques

24. The computer crime technique that involves planting instructions that eventually cause software to self-destruct is called ______.
    (a) super-zap
    (b) time bomb *
    (c) Trojan horse
    (d) trapdoor
    (e) sabotage

25. A computer crime technique frequently used by software developers to ensure payment for their work is the ______

(a) Trojan horse
(b) sabotage
(c) time bomb *
(d) super-zap
(e) trapdoor

26. With the ______ special instructions in programs can make unauthorized copies of other programs and data.
(a) time bomb
(b) trapdoor
(c) super-zap
(d) Trojan horse *
(e) salami method

27. The computer crime technique often used where interest is calculated is the ______ method.
(a) super-zap
(b) data diddling
(c) trapdoor
(d) scavenging
(e) salami *

28. Using the ______ method of computer crime, a financial institution's books will remain balanced even after the crime has been perpetrated.
(a) scavenging
(b) salami *
(c) super-zap
(d) trapdoor
(e) data diddling

29. When conducting ______, programs are checked to determine if they have been modified.
(a) credit checks
(b) surprise audits
(c) software audits *
(d) access codes
(e) ciphertexts

30. The ______ method of computer crime uses legitimate computer programs that systems programmers normally use to override system controls.
(a) super-zap *
(b) Trojan horse
(c) trapdoor
(d) time bomb
(e) data diddling

31. With the ______ method of computer crime, copies of a company's programs and data are obtained from memory dumps or even trash bins.

(a) data diddling
(b) Trojan horse
(c) time bomb
(d) scavenging *
(e) super-zap

32. Obtaining a memory dump from a company's computer is an example of the ______ method.
(a) super-zap
(b) Trojan horse
(c) salami
(d) data diddling
(e) scavenging *

33. With the ______ method of computer crime, information either entering or leaving the computer can be altered.
(a) Trojan horse
(b) data diddling *
(c) salami
(d) scavenging
(e) super-zap

34. With ______, computer programs are modified to alter information entering or leaving the computer.
(a) data diddling *
(b) scavenging
(c) program theft
(d) embezzlement
(e) sabotage

35. The ______ method of computer crime involves secret instructions that bypass system safeguards.
(a) data diddling
(b) salami
(c) trapdoor *
(d) scavenging
(e) super-zap

36. Using ______, students have been known to change their grades.
(a) stealing computer time
(b) sabotage
(c) embezzlement
(d) program and data theft or alteration *
(e) data diddling

37. The most common type of computer crime is ______.
(a) sabotage

(b) program and data theft or alteration
(c) data diddling
(d) scavenging
(e) stealing computer time *

38. _____ involves employees stealing money or goods from a company.
(a) Sabotage
(b) Embezzlement *
(c) Data diddling
(d) Trojan horse method
(e) Scavenging

39. The destroying of computers, data, and programs by such methods as fire or microwave signals is called _____.
(a) embezzlement
(b) scavenging
(c) sabotage *
(d) data diddling
(e) time bomb method

Case Study 15-1: Classic Computer Crimes

Slipping in Suspicious Slips

40. To avoid the depositing of money into the wrong accounts many banks require that tellers hand out _____.
(a) loan payment slips
(b) deposit slips *
(c) withdrawal slips
(d) credit slips
(e) passbooks

The Prep School Gang

41. To prevent unauthorized use of timesharing accounts, clients should be given _____.
(a) long passwords *
(b) pass keys
(c) signature verification keys
(d) voiceprints
(e) ciphertext

Diamonds Are a Bank's Best Friend

42. To prevent unauthorized electronic funds transfers banks should ______.
    (a) post their access codes
    (b) use encryption
    (c) use suprise audits
    (d) change access codes regularly *
    (e) use pass keys

The ATM Kidnap Caper

43. To prevent crimes at ATMs, many of these machines are equipped with ______.
    (a) pass keys
    (b) access codes
    (c) ciphertext
    (d) hand analyzers
    (e) photographic devices *

Funding Funds with Fake Funds

44. Many companies use ______ programs in order to prevent embezzlement by their employees.
    (a) software
    (b) time bomb
    (c) computer auditing *
    (d) Trojan horse
    (e) trapdoor

The Scavenger's Delight

45. A way to prevent important company correspondence or information from falling into the wrong hands is to use a ______.
    (a) backup copy
    (b) paper schredder *
    (c) software lock
    (d) access code
    (e) password

Stealing a Stolen System

46. For security purposes and also in case of theft it is wise for companies to have ______ of programs.
    (a) backup copies *

(b) decryption keys
(c) software audits
(d) signature verification
(e) software locks

The Prompt Prepayment Perpetrator

47. When banks issue loan payment books, it is important that their computers be programmed to ______
(a) issue the loan
(b) print the payment stubs
(c) keep track of the total number of payments *
(d) post interest
(e) withdraw the amount owed from savinngs accounts

The Computerized Bulletin Board Blitz

48. Long-distance phone companies are considering using ______ access codes to prevent fraud.
(a) encrypted
(b) plaintext
(c) longer *
(d) cyphertext
(e) decryption

Computer Criminals

49. Which of the following characteristics does not describe a typical computer criminal? ______
(a) Intelligent
(b) Has no previous criminal record
(c) Outwardly loyal to the company
(d) Fears detection more than punishment
(e) Over 30 years of age *

Computers and Law Enforcement

50. The center run by the federal government that keeps track of crimes and criminals is the ______
(a) NCEFT
(b) NCIC *
(c) EFT
(d) POS

(e) FBI

Security

51. Security devices that resemble credit cards are ______.
    (a) passkeys *
    (b) signature verifications
    (c) access codes
    (d) cyphergrams
    (e) hand analyzers

52. To improve security many companies commonly use both passkeys and ______.
    (a) hand analyzers
    (b) voiceprints
    (c) access codes *
    (d) fingerprints
    (e) signature verification

Disaster Recovery

53. A device that prevents the loss of power in a computer installation is a ______.
    (a) POS
    (b) ATM
    (c) NCEFT
    (d) UPS *
    (e) NCIC

Encryption

54. The coding of information is called ______.
    (a) decryption
    (b) passkeys
    (c) verification
    (d) ciphertext
    (e) encryption *

55. A ______ is used to decode encoded messages.
    (a) ciphertext
    (b) decryption key *
    (c) plaintext
    (d) software lock
    (e) passkey

56. An encrypted message is called a ______.
    (a) cryptogram *
    (b) plaintext
    (c) cleartext
    (d) access code
    (e) software lock

Software Protection

57. The use of ______ prevents the unauthorized copying of computer programs.
    (a) ciphertext
    (b) encryption
    (c) software locks *
    (d) access codes
    (e) backup copies

Privacy

58. The ______ gives individuals the right to know the operation and structure of federal agencies, and which ones keep records on them.
    (a) Privacy Act
    (b) Fair Credit Reporting Act
    (c) Right to Financial Privacy Act
    (d) Freedom of Information Act *
    (e) Education Privacy Act

59. Under the ______, credit agencies may collect only that information which is needed to verify a person's credit.
    (a) Fair Credit Reporting Act *
    (b) Electronic Funds Transfer Act
    (c) Privacy Act
    (d) Right to Financial Privacy Act
    (e) Freedom of Information Act

60. The law that applies when someone wishes to find out why they were denied credit is the ______.
    (a) Privacy Act
    (b) Fair Credit Reporting Act *
    (c) Electronic Funds Transfer Act
    (d) Right to Financial Privacy Acy
    (e) Freedom of Information Act

61. The ______ applies to academic institutions that receive federal funds.
    (a) Freedom of Information Act

(b) Fair Credit Reporting Act
(c) Education Privacy Act *
(d) Electronic Funds Transfer Act
(e) Right to Financial Privacy Act

62. The ______ prevents the federal government from keeping secret files on individuals.
(a) Right to Financial Privacy Act
(b) Education Privacy Acy
(c) Freedom of Information Act
(d) Privacy Act *
(e) Fair Credit Reporting Act

63. The ______ gives individuals the right to review data about their personal finances.
(a) Electronic Funds Transfer Act
(b) Privacy Act
(c) Freedom of Information Act
(d) Fair Credit Reporting Act
(e) Right to Financial Privacy Act *

True/False

Introduction

1. With EFT, money is transferred electronically without the use of written checks. (T p398)

2. Preauthorized computer-initiated payments is an effective way to handle a family's weekly purchases at the supermarket. (F p398)

3. It is clear that we are headed toward a cashless/checkless society. (F p398)

4. Today's view of EFT has been tempered somewhat to the less cash/less check society. (T p398)

5. Today, the vast majority of banking is done electronically. (F p398)

Direct Deposit

6. One of the earliest forms of EFT was direct deposit of payroll involving a single company and a single bank. (T p398)

7. Electronic funds transfers between banks are handled through a series of computer centers called automated clearing houses run by the NCEFT. (F p399)

8. The Social Security Administration has strongly encouraged participants to use direct deposit programs. (T p399)

9. Direct deposit is an important convenience to many elderly and disabled people. (T p399)

On-Line Teller Terminals

10. With on-line teller terminals, bank tellers have instant access to customer accounts. (Tp 399)

11. On-line teller terminals speed the process of verifying balances before allowing cash withdrawls. (T p399)

Automated Teller Machines

12. With ATMs, banks can offer many teller services around the clock. (T p399)

13. ATMs are generally off-line from the bank's computers. (F p400)

14. A person must have a PIN to use an ATM in EFT. (T p400)

15. ATMs have allowed many banks to cut back their regular banking hours. (T p400)

16. ATM transactions are more costly to the banks than teller transactions, but the banks keep using the ATMs for customer convenience. (F p401)

17. ATM users should write their PIN numbers on their ATM cards because if they forget their PINs they won't be able to get the ATM to work. (F p401)

18. Regardless of when you notify the bank regarding losses on your account (possibly due to a stolen ATM card), government rules and regulations require that the banks absorb all ATM losses; this is an example of consumer protection legislation. (F p401)

Credit Authorization Terminal

19. Credit authorization terminals are normally off-line to computers that maintain customer credit information. (F p402)

The Debit Card and the Credit Card

20. When a client uses a debit card, that client is essentially borrowing money to pay for the item being purchased. (F p402)

21. When a client uses a debit card to buy something, the funds are transferred over the EFT networks from the store's bank account to the client's account immediately. (F p402)

22. A debit card can be used like cash to pay for items directly. (T p402)

23. As debit cards become more widely used, they may eventually eliminate the usefulness of traveler's checks. (T p402)

Pay-by-Phone

24. One problem with EFT is that there is no way to implement an electronic stop payment order for the client who wants to hold up a payment. (F p402)

25. Many banks offer pay-by-phone services to customers with touch tone phones; the phones actually act as digital data entry devices. (T p402)

26. Because banks are supplying an increasing number of services to bank-at-home customers, it is possible that they may eventually become key videotex providers. (T p402)

Point-of-Sale Terminals and Transactions

27. POS systems facilitate checking out customers in supermarkets and other retail establishments. (T p403)

28. OMR is a key technology in today's POS systems. (F p403)

Privacy and Social Issues in EFT

29. Someone with access to EFT records could conceivably trace another person's spending habits and movements, so the FBI has tapped into all the major EFT networks. (F p403)

30. Welfare agencies around the United States use EFT information to monitor the purchases of welfare recipients to be sure they are confined to basic necessities. (F p404)

31. If only summary information about EFT transactions is made available to businesses and government agencies, then there is no danger to the invasion of individual privacy. (F p404)

EFT Legislation and Regulation

32. The NCEFT is the government agency that controls EFT throughout the United States.

(F p405)

33. The EFT Act of 1978 contained the recommendations of a group appointed by Congress to specify how EFT systems should be monitored and controlled. (T p405)

34. The NCEFT recommended that EFT be developed and controlled primarily by the federal government. (F p405)

35. The NCEFT recommended that customers be fully liable for lost, stolen, or misused debit cards. (T p405)

36. One interesting recommendation of the NCEFT was that at the time of sale the merchant should not be able to determine if the customer is paying by a credit card or a debit card. (T p405)

37. Perhaps the most significant set of rules governing EFT is Regulation E of the Federal Banking Commission. (F p403)

38. Regulation E provides that if a bank's dispute over a transaction with a customer cannot be resolved immediately, then the funds in question are frozen until the dispute can be resolved. (F p405)

Computer Crime Techniques

39. The key to distinguishing computer crimes from non-computer crimes is that computer crimes cannot be performed without the use of computers. (F p406)

40. The "time bomb" is a computer crime technique frequently used by people who develop software; if they are not paid within a reasonable period of time, the software literally destroys itself. (T p406)

41. Beware of borrowed software; as it executes, it may make copies of your programs and data for the "nice" person who lent it to you; this is an example of the Trojan horse method of computer crime. (T p406)

42. With the scavenging method of computer crime, the fractions of a cent that occur when a bank calculates interest on customer accounts are placed in the account of the thief. (F p406)

43. Periodic software audits are more effective than supprise audits in catching computer criminals. (F p406)

44. With the super-zap method of computer crime, legitimate computer programs ordinarily used by systems programmers to override normal system controls and modify data are used for criminal purposes. (T p406)

45. Running a memory "dump" to obtain a copy of all the programs and data in the computer is an example of the trapdoor method of computer crime. (F p406)

46. One of the simplest forms of computer crime is the scavenging method in which someone rummages through the trash for computer printouts and documentation. (T p406)

47. The data diddling technique of computer crime involves modifying a computer program so that information either entering or leaving the system can be changed. (T p406)

48. With the trapdoor method of computer crime, secret instructions are placed in target programs to enable the thief to bypass system safeguards whenever the thief wishes to use the programs. (T p407)

49. The Department of Defense developed a completely secure computer system that can withstand any possible attack by highly skilled users. (F p406)

50. The "take" in computer-assisted embezzlements is often 20 times as great as in manual embezzlements. (T p407)

Case Study 15-1: Classic Computer Crimes

Slipping in Suspicious Slips

51. Banks should not allow customers to make more than one deposit per day. (F p407)

52. Bank computers should highlight excessive deposit activity on any one account to prevent certain types of computer crime. (T 407)

The Prep School Gang

53. Timesharing system passwords should be three digits long to prevent people from penetrating the system by trying all possible passwords. (F p407)

54. Timesharing systems should monitor excessive attempts to login with invalid passwords and should inform administrators who could in turn inform the police. (T p407)

55. To discourage attempts to try all possible passwords, some timesharing systems automatically hang up on a user who has typed three consecutive invalid passwords. (T p407)

Diamonds Are a Bank's Best Friend

56. Passwords and secret access codes used to gain access to computer systems should be changed frequently so that stolen codes will be useful for only a short time. (T p408)

The ATM Kidnap Caper

57. Legislation has placed the burden of proof in most disputes involving ATM transactions on the consumer rather than the bank. (F p408)

58. Legislation has forbidden banks from photographing ATM users because this is considered to be an invasion of privacy. (F p408)

Funding Funds with Fake Funds

59. It is clear that auditing computer systems manually is sufficient to uncover all computer frauds. (F p408)

60. Computers should not be used to audit other computers because the audit computers themselves could be the object of computer crimes. (F p408)

The Scavenger's Delight

61. Paper shredders prevent any scavenging-type computer crimes. (F p409)

62. Convicted computer criminals often receive light sentences. (T p409)

Stealing a Stolen System

63. Unlike stealing merchandise, it is possible to steal computer programs and data by making copies and leaving the originals intact. (T p409)

The Prompt Prepayment Perpetrator

64. In general, computers cannot overcome programmer or management stupidity in poorly designed or poorly managed systems. (T p409)

The Computerized Bulletin Board Blitz

65. Automatic dialing modems are a boon to computer criminals. (T p409)

66. Posting other people's long-distance telephone codes on computerized bulletin boards is not a

crime, but using those codes to charge calls to other people is. (T p410)

67. Long-distance telephone companies and their clients alike agree that very long secret access codes should be used to thwart criminals who discover codes by trying all possibilities. (F p410)

68. Stiffening the laws and penalties for using computers to perpetrate crimes would discourage more potential computer criminals. (T p410)

Computer Criminals

69. According to the FBI, computer criminals are generally over 30 years of age. (F p410)

70. Many companies would rather suffer their losses than admit that they have been victimized by computer criminals. (T p410)

71. Judges generally give convicted computer criminals stiff sentences to discourage potential computer criminals. (F p410)

Computers and Law Enforcement

72. The NCIC run by the federal government is a network of computer systems designed to track the whereabouts of every American. (F p411)

73. One of the real benefits of NCIC is that police officers are able to determine if the suspects they are dealing with are dangerous criminals. (T p411)

Security

74. A passkey is an effective means of identifying the person holding the card. (F p412)

75. Security systems that require both a passkey and a secret access code are not that much more effective than passkey only systems in identifying the person holding the card. (F p412)

76. A security system called the hand analyzer identifies people by their hand geometry. (T p412)

Disaster Recovery

77. Organizations must ensure that their computers will continue to operate in the wake of every conceivable disaster. (F p413)

78. One of the most effective recovery techniques is to keep backup copies of all programs and data locked in fireproof vaults, preferably at an alternative site. (T p413)

79. One of the most common problems faced by computer centers is loss of power. (T p413)

80. A computer center that must continue operating even during power failures may use a hardware device called a UPS. (T p413)

Encryption

81. Encryption is an effective security technique for preventing even stolen data from being of any use. (T p413)

82. An unencrypted message is called ciphertext. (F p413)

Software Protection

83. It is a crime to make any copies of copyrighted software. (F p414)

84. No one has ever invented a completely effective method of preventing software piracy. (T p414)

85. It is expected that the solution to the software piracy problem will come as soon as competition forces the price of software so low that it will be more economical to buy it than to copy it. (F p414)

Privacy

86. The Federal Agencies Act of 1966 gives individuals the right to know the operation and structure of federal agencies and which federal agencies keep records on them (F p415)

87. The Fair Credit Reporting Act of 1970 was passed because of the increasing collection of credit information made possible by computers and data communications. (T p415)

88. The Fair Credit Reporting Act of 1970 limits credit agencies to collecting only that information needed to verify a person's credit. (T p415)

89. The Education Privacy Act of 1974 offers privacy protection to students at all American colleges and universities. (F p415)

90. The Privacy Act of 1974 prevents the federal government from keeping secret files on individuals. (T p415)

91. The Privacy Act of 1974 forbids federal agencies such as the FBI and the IRS from sharing information they collect. (F p415)

92. The Right to Financial Privacy Act of 1978 gives people the right to review the data about their personal finances that financial institutions maintain about them. (T p415)

93. The increasing use of home computers and two-way cable TV will help secure our privacy. (F p416)

Fill-in-the-Blanks

1. **Electronic funds transfer (EFT)** is the process by which money is transferred electronically by computers and data communications systems without writing checks.

Direct Deposit

2. One of the earliest forms of EFT was **direct deposit** of payroll involving a single company and a single bank.

3. Electronic fund transfers between banks are handled through a series of computer centers called **automated clearing houses** run by the Federal Reserve Board.

On-Line Teller Terminals

4. Today, most banks use **on-line** teller terminals so that information can be entered directly into their computers.

Automated Teller Machines

5. With the use of **automated teller machines (ATMs)** banks can now offer many teller services around the clock.

6. To operate an ATM, a client needs an identification card, a **personal identification number** (PIN) known only to the client and the bank, and an account at the bank.

Credit Authorization Terminals

7. **Credit authorization** terminals are used by retail merchants to help determine if checks and credit cards are valid.

Enrichment Piece: Using ATMs

8. When using an ATM it is important not to write your **PIN** on your identification card.

9. When using an ATM it is a good idea to insist on **privacy** while using ypur identification card.

The Debit Card and The Credit Card

10. When a person uses a **credit** card they are essentially borrowing money to pay for their purchase.

11. When a person uses a **debit** card, funds are transferred over EFT networks from the person's account to the account of the store.

Pay-by-Phone

12. Many banks offer **pay-by-** phone services to clients with touch-tone phones.

13. For customers with personal computers, many banks offer **bank-at-home** services that are more elaborate than pay-by-phone services.

Point-of-Sale Terminals and Transactions

14. **Point-of-sale** (POS) terminals facilitate retail transactions in stores, supermarkets, and other places of business.

Privacy and Social Issues in EFT

EFT Legislation and Regulation

15. The commission created by Congress to recommend legislation that would specify how EFT systems should be monitored and controlled was the **National Commission on Electronic Fund Transfers (NCEFT)**.

16. Congress incorporated the recommendations of the NCEFT into the **Electronic Funds Transfer** Act.

17. **Regulation E (Reg E)** implements the provisions of the Electronic Funds Transfer Act

and provides considerable protection to customers.

The Future of EFT

Computer Crime Techniques

18. The computer crime technique used by software developers that will make software destruct if a software developer is not paid by a certain time is the **time bomb**.

19. The **Trojan hourse method** of computer crime involves placing special instructions into the computer program; the instructions perform secret functions while the program appears to run normally.

20. The method of computer crime that involves taking small amounts from many different accounts and placing these amounts in the account of the thief is the **salami method**.

21. **Software** audits carefully examine computer programs to determine if they have been secretly modified.

22. **Surprise** audits are important because they help catch computer criminals off guard.

23. The computer crime technique that involves making improper use of a legitimate computer program ordinarily used by systems programmers is the **super-zap** method.

24. The **scavenging** method can be performed by either rummaging through trash for computer printouts or running a memory "dump" to obtain computer programs.

25. **Data diddling** is a computer crime that involves modifying a computer program so that information either entering or leaving the system can be changed.

26. With the **trapdoor** method of computer crime, secret instructions are placed in target programs to enable the criminal to bypass system safeguards and use these programs at will.

27. **Program and data theft or alteration** involves the theft or modification of either programs or data for the person's own benefit.

28. The most common type of computer crime is **stealing computer time**.

29. **Embezzlement** is a type of computer crime tin which an employee steals money or goods from a company.

30. The type of crime that involves the destroying of computers, data, and programs by using fire, bombing, and turning off air conditioning or electrical systems is **sabotage**.

Computer Criminals

31. According to the FBI, computer criminals are generally under **30** years of age.

Computers and Law Enforcement

32. Perhaps the most important application of computers in law enforcement today is the **National Crime Information Center (NCIC)** run by the federal government.

Security

33. A security device that resembles a credit card is a **passkey** .

34. A multi-digit secret **access code** is often used with a passkey to improve security.

35. A security system called the **hand analyzer** identifies people by their hand geometry.

36. **Voiceprint** identification systems match people's voices to prestored voice samples.

37. **Signature verfication** systems match signatures to prestored samples; they can even measure changes in pressure as people write.

Disaster Recovery

38. A computer installation that must continue operating even during power failures may use a hardware device called an **uninterruptible power supply (UPS)** .

Encryption

39. To render data useless to unauthorized users, it may be **encrypted** (coded).

40. To decode an encrypted message requires knowledge of a **decryption** key.

41. An encrypted message is called a cryptogram or **ciphertext** .

42. Unencrypted messages are called plaintext or **cleartext** .

Software Protection

43. The developers of programs used on personal computers sometimes build in **software locks** that either prevent duplication or allow a buyer to make a limited number of backup copies.

Privacy

44. The **Freedom of Information** Act gives individuals the right to know the operation and structure of federal agencies and which federal agencies keep records on them.

45. The **Fair Credit Reporting** Act gives individuals the right to determine what credit information is being collected about them and to force the correction of inaccurate information.

46. The **Education Privacy** Act ensures the privacy and security of academic and behavioral records and permits access to a student's records only by that student, the student's parents, or authorized school officals.

47. The **Privacy** Act prevents the federal government from keeping secret files on individuals, and prevents government agencies from sharing information without the authorization of a federal judge.

48. The **Right to Financial Privacy** Act gives people the right to review the data about their personal finances that financial institutions maintain about them.

Chapter 16
Robotics and Artificial Intelligence

Multiple Choice

1. The study of making machines exhibit intelligent behavior is called ______.
   (a) expert systems
   (b) natural language processing
   (c) knowledge structures
   (d) artificial intelligence *
   (e) robotics

Robotics

2. Most of today's ______ are relatively dumb machines.
   (a) knowledge engineers
   (b) programmers
   (c) robots *
   (d) domain experts
   (e) politicians

3. Because of the high costs of labor, ______ can perform many of the same tasks for less money.
   (a) industrial robots *
   (b) audio-animatrons
   (c) expert systems
   (d) anthropomorphic robots
   (e) domain experts

The Mechanical Human

4. Robots that have the form and attributes of humans are called ______ robots.
   (a) assembly
   (b) intelligent
   (c) sensory
   (d) computerized
   (e) anthropomorphic *

5. The electromechanical people and animals created by the Disney imagineers are called ______.
   (a) sensory robots
   (b) audio-animatrons *
   (c) computerized robots
   (d) anthropomorphic robots
   (e) flexible robots

6. ______ robots are being used as teaching tools in many medical schools.
   (a) Anthropomorphic *
   (b) Sensory
   (c) Flexible
   (d) Audio-animatron
   (e) Programmable

A Brief History of Robotics

7. The word robot was coined by ______.
   (a) Joseph Engelberger
   (b) Isaac Asimov
   (c) Neale Clapp
   (d) Karel Capek *
   (e) Alan Turing

8. ______ wrote the play *R. U. R.*, in which humaniod machines ultimately destroy humanity.
   (a) Marvin Minsky
   (b) Neale Clapp
   (c) Karel Capek *
   (d) Isaac Asimov
   (e) Joseph Weizenbaum

9. The father of industrial robots is ______.
   (a) Joseph Weizenbaum
   (b) Isaac Asimov
   (c) Neale Clapp
   (d) Marvin Minsky
   (e) Joseph Engelberger *

10. Which of the following characteristics applies to first generation robots? ______
    (a) elementary vision and touch
    (b) dumb manipulators *
    (c) limited intelligence
    (d) highly intelligent
    (e) equipped with hands

11. Of the following characteristics ______ does not apply to second generation robots.
   (a) touch
   (b) arms
   (c) skin *
   (d) optic sensors
   (e) hands

12. When the fourth generation of robots appears they will have ______.
   (a) elementory vision and touch
   (b) limited intelligence
   (c) optic sensors
   (d) high intelligence *
   (e) dumb manipulators

Productivity Concerns

13. The country that uses the most robots is ______.
   (a) Japan *
   (b) Russia
   (c) West Germany
   (d) United States
   (e) France

14. Changing assembly lines adjust quickly to producing different products is called ______
   (a) expert systems
   (b) flexible automation *
   (c) bin picking
   (d) findustrial automation
   (e) batch processing

Industrial Robots: "Steel Collar" Workers

15. ______ robots are used for welding, spray painting, and manipulating nuclear fuel rods.
   (a) Pick-and-place
   (b) Intelligent
   (c) Industrial *
   (d) Intuitive
   (e) Bin picking

16. Of the following countries ______ is considered to be 10 years behind the rest in the use of robots.
   (a) France
   (b) Japan
   (c) United States

(d) Russia *
(e) Italy

How Robots Work

17. The robots that select an object from one area and put them in another are referred to as ______ robots.
    (a) programmable
    (b) bin picking
    (c) sensory
    (d) assembly
    (e) pick-and-place *

18. ______ robots repeatedly perform only one task sequence on command.
    (a) Assembly
    (b) Pick-and place *
    (c) Sensory
    (d) Programmable
    (e) Computerized

19. With ______ robots a, person leads the robot's arm amnually through the series of movements and tasks it is to perform.
    (a) programmable *
    (b) assembly
    (c) pick-and-place
    (d) sensory
    (e) computerized

20. ______ robots are designed to remember movements demonstrated by human "teachers," and faithfully repeat the sequences upon command.
    (a) Computerized
    (b) Pick-and-place
    (c) Programmable *
    (d) Assembly
    (e) Sensory

21. The reprogramming of ______ robots is a tedious manual process.
    (a) sensory
    (b) assembly
    (c) pick-and-place
    (d) computerized
    (e) programmable *

22. ______ robots are software programmable.
    (a) Industrial

(b) Programmable
(c) Pick-and-place
(d) Computerized *
(e) Anthropomorphic

23. With ______ robots instructions are transmitted electronically from a computer to the robot.
(a) computerized *
(b) industrial
(c) programmable
(d) pick-and-place
(e)anthropormorphic

24. ______ robots can be reprogrammed almost instantly.
(a) Industrial
(b) Computerized *
(c) Anthropomorphic
(d) Pick-and-place
(e) Programmable

25. ______ robots have limited touch or vision capabilities.
(a) Programmable
(b) Pick-and-place
(c) Assembly
(d) Computerized
(e) Sensory *

26. Once robot touch and vision capabilities are sufficently developed ______ robots will be put to use putting items together.
(a) sensory
(b) assembly *
(c) pick-and-place
(d) computerized
(e) programmable

27. The ability of robots to distinguish between many intermixed parts has been give the name of the ______ problem.
(a) sensory
(b) part picking
(c) bin picking *
(d) domain
(e) flexible automation

Robot Vision

28. The next major advance in robots is considered to be ______.
    (a) vision *
    (b) touch
    (c) smell
    (d) hearing
    (e) taste

An Artificial Sense of Touch

29. To enable robots to have a sense of touch, scientists at MIT have developed robotic ______.
    (a) sensors
    (b) hands
    (c) fingers
    (d) skin *
    (e) expert systems

Household Robots

30. The most complex area for a robot to function in is a ______.
    (a) factory
    (b) assembly line
    (c) laboratory
    (d) industrial plant
    (e) house *

Organized Labor and Robotics

31. Of the following items, ______ does not describe what type of job robots now hold in United States industry.
    (a) boring
    (b) dirty
    (c) monotonous
    (d) creative *
    (e) hazardous

Box: Clapp's Laws of Robots

32. Which of the following was not suggested by Neal Clapp? ______
    (a) robots cannot be installed by devious strategies
    (b) robots should not be installed to replace human labor *
    (c) robots should be installed where humans are performing like robots
    (d) robots should not be installed to the economic detriment of workers

(e) robots shall not be installed through distrust

Philosophical Issues

33. Of the following capabilities that future robots may have, ______ is considered by many to be the most frightening.
    (a) producing future members of their species that are better than current members *
    (b) reasoning faster
    (c) having great strength
    (d) having huge memory capacity
    (e) being able to swim and fly

34. With the field of ______, we might eventually be able to build living creatures to meet our own specifications.
    (a) expert systems
    (b) explanation systems
    (c) knowledge engineering
    (d) domain experts
    (e) genetic engineering *

The Turing Test of Intelligence

35. The standard test used to test for machine intelligence was developed by ______.
    (a) Joseph Weizenbaum
    (b) Marvin Minsky
    (c) Alan Turing *
    (d) Neale Clapp
    (e) Joseph Engleberger

36. The ______ Test is used to determine if a machine is exhibiting intelligent behavior.
    (a) Explanation
    (b) Knowledge
    (c) Domain
    (d) Turing *
    (e) Heuristic

Expert Systems

37. ______ are computer systems that can provide intelligent answers to questions in specialized areas.
    (a) Knowledge bases
    (b) Expert systems *
    (c) Domains

(d) Heuristics
(e) Explanation systems

38. Tax advice, medical diagnoses, and evaluating prospecting information are capabilities of ______ systems.
(a) expert *
(b) MIS
(c) DSS
(d) DAS
(e) EDP

Expert System Components

39. The component of an expert system that contains the rules of thumb for a given area of interest is the ______.
(a) explanation system
(b) workspace
(c) inference engine
(d) natural language processor
(e) knowledge base *

40. Another name for the knowledge base is ______.
(a) workspace
(b) explanation
(c) domain *
(d) inference engine
(e) natural language processor

41. The ______ is a set of programs that reasons from the information contained in the knowledge base.
(a) domain expert
(b) inference engine *
(c) natural language processor
(d) explanation system
(e) workspace

42. The component of an expert system that allows the user to carry on a dialogue with the computer is the ______.
(a) workspace
(b) domain expert
(c) inference engine
(d) natural language processor *
(e) explanation system

43. The ______ in an expert system carries on a user-friendly dialogue.

(a) natural language processor *
(b) explanation system
(c) domain expert
(d) knowledge engineer
(e) workspace

44. The component of an expert system that the computer uses as a scratch pad is the ______.
(a) explanation system
(b) domain
(c) workspace *
(d) natural language processor
(e) inference engine

45. ______ is an area of memory used by an expert system in which the user's problem statement is placed.
(a) Inference engine
(b) Domain
(c) Natrual language processor
(d) Explanation system
(e) Workspace *

46. The ______ component of an expert system details for the user how conclusions were reached.
(a) inference engine
(b) explanation system *
(c) domain
(d) workspace
(e) knowledge base

47. The ______ obtains information for the knowledge base.
(a) domain
(b) domain expert
(c) natural language translation
(d) knowledge engineer *
(e) inference engine

48. A human expert in a particular field is also called a ______.
(a) domain expert *
(b) knowledge base
(c) knowledge engineer
(d) natural language translator
(e) inference engine

Natural Language Processing

49. _____ is concerned with the problems of developing machines that understand languages like English, Spanish, and French.
   (a) Ambiguity
   (b) Natural language processing *
   (c) Knowledge structure
   (d) Speech synthesis
   (e) VHLL

50. The term _____ describes the converting of English into French.
   (a) serial processor
   (b) knowledge structures
   (c) natural language translation *
   (d) ambiguity
   (e) speech synthesis

51. Because of _____ there can never be perfect translation.
   (a) natural language processing
   (b) serial processors
   (c) natrual language translation
   (d) knowledge structures
   (e) ambiguity *

Heuristics

52. A _____ is a rule of good judgment based on experience.
   (a) heuristic *
   (b) knowledge structure
   (c) domain
   (d) explanation system
   (e) knowledge expert

53. A _____ gives an opponent the smallest number of moves.
   (a) knowledge structure
   (b) fewest-replies heuristic *
   (c) domain
   (d) serial processor
   (e) knowledge base

The Game of Chess

54. The game of _____ has received the most interest among artificial intelligence researchers.
   (a) checkers
   (b) backgammon

(c) chess *
(d) craps
(e) cribbage

55. Artificial intelligence researchers have generally viewed as an extremely significant achievement, a computer that could beat the world's best ______ player.
(a) craps
(b) cribbage
(c) checkers
(d) chess *
(e) backgammon

Box: Eliza and the Turing Test

56. The computer system that acted the part of a phychiatrist was ______.
(a) LISP
(b) Turing
(c) Unimation
(d) RUR
(e) Eliza *

57. The Eliza program was developed by ______.
(a) Alan Turing
(b) Joseph Weizenbaum *
(c) Marvin Minsky
(d) Joseph Engleberger
(e) Karel Capek

LISP: The Programming Language of Artificial Intelligence

58. The programming language that primarily manipulates symbols is ______.
(a) LISP *
(b) COBOL
(c) Pascal
(d) BASIC
(e) Fortran

59. ______ are examples of knowledge structures.
(a) Serial processors
(b) Multiprocessors
(c) If-then rules *
(d) Domains
(e) Heuristics

Serial Versus Parallel Computers

60. Today's processors are ______, that process one task at a time.
    (a) parallel processors
    (b) multiprocessors
    (c) massive processors
    (d) serial processors *
    (e) natural language processors

61. Computers that have several processors are called ______.
    (a) natural language processors
    (b) serial processors
    (c) massive processors
    (d) uniprocessors
    (e) multiprocessors *

62. The term used to describe computers that have hundreds and even thousands of processors is ______.
    (a) multiprocessing
    (b) massive parallelism *
    (c) natural language processing
    (d) serial processing
    (e) parallel processing

True/False

Introduction

1. Artificial intelligence is the study of making intelligent machines. (F p424)

2. Artificial intelligence is generally considered to be the most controversial field in computer science. (T p424)

Robotics

3. Most of today's robots are the familiar steel people from science fiction, movies, and television. (F p424)

4. Most of today's robots are relatively dumb machines. (T p424)

The Mechanical Human

5. Anthropomorphic robots have the form and attributes of human's. (T p426)

6. Disney calls its electromechanical people and animals "Imagineers." (F p426)

A Brief History of Robotics

7. The word robot comes from the Check noun "robota" meaning "forced labor." (T p427)

8. *R.U.R* was written by Isaac Asimov. (F p427)

9. The first production line robot was built by General Motors. (F p427)

10. Joseph Engelberger is generally recognized as the father of industrial robotics. (T p427)

11. The first generation robots in wide use today are little more than dumb mechanical manipulators. (T p427)

12. Second generation robots have elementary sensory capabilities. (T p427)

13. Third generation robots (late 1980s) will have highly developed intelligence. (F p427)

Productivity Concerns

14. The United States is by far the largest user of industrial automation and robots in the world. (F p428)

15. The vast majority of goods made in America are not mass produced. (T p428)

16. Adaptable automation used in small batch environments is called flexible automation. (T p428)

Asimov's Laws of Robotics

17. The term "robotics" was coined by Karel Capek. (F p428)

18. Isaac Asimov is considered to be the father of fictional robotics. (T p428)

Industrial Robots: "Steel Collar" Workers

19. Industrial robotics in the United States was a $100 billion industry in 1980. (F p429)

20. The main drawback of today's industrial robots is their high initial cost. (T p429)

21. It is interesting to note that Communist countries are several years ahead of the West and Japan in their use of robots. (F p429)

How Robots Work

22. Most robots in use today are one-armed industrial robots used to perform various tasks on highly automated assembly lines. (T p429)

23. Pitch, yaw, roll, elbow extension, shoulder swivel, and arm sweep are all typical parts on today's industrial robots. (F p429)

24. Pick-and-place robots are among the most sophisticated of today's industrial robots. (F p429)

25. Pick-and-place robots repeatedly perform only one task sequence on command. (T p429)

26. Programming a robot is not an easy task--it has been likened to trying to write down how to tie a shoe. (T p429)

27. With programmable robots, a person leads the robot's arm manually through the series of movements and tasks it is to perform; the robot remembers these and repeats them on command. (T p429)

28. Computerized robots are software programmable. (T p430)

29. The key problem that must be solved before assembly robots will be really effective is called the bin picking problem. (T p430)

Robot Vision

30. It is generally believed that the next major advance in robotics will be the use of computerized vision systems that will enable robots to "see." (T p430)

31. Robot vision systems that rival human vision are expected to appear within a year or two. (F p430)

32. The pictures seen by today's robots normally contain much more detail than humans are able to perceive. (F p430)

33. One common application for today's robots with vision is as parts inspectors an assembly lines. (T p430)

An Artificial Sense of Touch

34. Researchers at M.I.T. have developed a robotic skin. (T p432)

35. Today's industrial robots use crude gripper mechanisms to lift objects. (T p432)

Household Robots

36. An assembly line is a far more complex place for a robot than a home. (F p433)

37. U.S. toy manufacturers are reputed to be spending more money developing robot toys than NASA spends developing space exploration robots. (T p433)

Organized Labor and Robotics

38. Unions generally have not been as violently opposed to robots as might be expected. (T p433)

39. The robotics industry in the United States is large and growing quickly. (F p433)

40. Robots are already replacing large numbers of white-collar workers. (F p433)

41. Companies have generally avoided firing workers whose jobs were eliminated by robots. (T p433)

42. The loss of the United States's competitive edge in world markets poses a much greater threat to blue-collar workers than robots do. (T p434)

43. One factor that encourages automation is the labor gap that is increasing as fewer people enter the manufacturing trades each year. (T p434)

44. The growing robotics industry is creating many new jobs each year. (T p434)

45. The countries with the greatest numbers of robots tend to have some of the highest unemployment rates. (F p434)

46. A major issue of concern to unions dealing with automation of their industries is job security for future generations of workers. (F p434)

Clapp's Laws of Robotics

47. "Organizations may not install robots to the economic, social, or physical detriment of

workers or management." is one of Asimov's Laws of Robotics. (F p434)

48. "A robot may not injure a human being, or, through inaction, allow a human being to come to harm." is one of Clapp's Laws of Robotics. (F p434)

49. "Organizations may only install robots on those tasks which, while currently performed by humans, are tasks where the human is like a robot, not the robot like a human." is one of Clapp's Laws of Robotics. (T p434)

Elimination of the Work Ethic

50. It is possible that robots may eventually eliminate the need for people to work for a living. (T p435)

What Will People Do?

51. If robots eliminate the need for humans to work, then humans will not work at all. (F p436)

Who Will Control the Wealth?

52. It is clear that in a heavily robotized society, the government should control all robots and the wealth they produce. (F p436)

Philosophical Issues

53. It is clear that a great deal of thought needs to go into considering what a heavily robotized society might be like. (T p437)

The Turing Test of Intelligence

54. The Asimov Test is considered the standard test of machine intelligence. (F p438)

55. Alan Turing stated that if an interrogator asking questions of a person and a machine couldn't tell by the answers alone which was the person, then the machine was exhibiting intelligence. (T p438)

Expert Systems

56. Expert systems "immortalize" expertise. (T p438)

57. Expert systems may be duplicated easily so that the expertise may be used in many places simultaneously. (T p438)

58. Today many humans already blindly obey machines because they are told to. (T p439)

Expert System Components

59. The knowledge base of an expert system contains the factual knowledge and the rules of thumb of a given area of interest. (T p439)

60. The area of interest of an expert system is called the range of interest. (F p439)

61. In an expert system, the set of programs that reasons from the information contained in the knowledge base is called the implication engine. (F p439)

62. In an expert system, the natural language processor is a set of programs that enables the user and the computer to carry on a user-friendly dialogue. (T p439)

63. An important component of an expert system is the explanation system that details for the user precisely how each conclusion is reached. (T p439)

64. In most envisioned applications, the expert system acts only as a consultant--not as a final authority. (T p439)

65. In expert systems development, the domain expert painstakingly interviews the knowledge engineer to obtain information for the knowledge base. (F p439)

Natural Language Processing

66. The problem of natural language translation is to build a device that will translate text from one language to another. (T p439)

67. It is widely believed that perfect natural language translation will be possible in the next century. (F p439)

68. One "sticky problem" in natural language translation is ambiguity. (T p439)

69. "The red brick factory manufactures red bricks" is a true statement. (F p439)

70. "The red brick factory is a factory built of red bricks" is a true statement. (F p439)

71. "The red brick factory is a building in which bricks are manufactured" is a true statement. (F p439)

Heuristics

72. Heuristics rarely guarantee success. (T p440)

73. In chess, "make the move that gives your opponent the smallest number of return moves" is an example of a heuristic. (T p440)

Eliza and the Turing Test

74. Perhaps the most famous natural language-based system was the Eliza program developed in the early 1960s at MIT by Joseph Weizenbaum. (T p440)

75. Weizenbaum's Eliza program imitated a psychiatrist. (T p440)

76. Weizenbaum's Eliza program could understand English sentences typed by the user. (F p440)

77. Joseph Weizenbaum was delighted that psychiatrists felt Eliza-like programs might eventually provide psychiatric services for humans. (F p440)

78. In spite of its ability to simulate an English dialogue, no one feels that Weizenbaum's Eliza program ever passed the Turing Test of intelligence. (F p440)

The Game of Chess

79. Chess playing machines that play a perfect game of chess are expected to become available in the 1990s. (F p441)

80. One key goal of artificial intelligence researchers has been to create a chess playing computer that could beat the world's best human player. (T p441)

81. Most researchers agree that a machine that could play chess better than any human would necessarily be intelligent. (F p441)

Lisp: The Programming Language of Artificial Intelligence

82. LISP is the preferred programming language for artificial intelligence applications. (T p441)

83. Unlike most programming languages that manipulate symbols, Lisp manipulates numbers. (F p441)

84. A Lisp program may process itself. (T p442)

Serial Versus Parallel Computers

85. Most computer systems today use multiple CPUs. (F p442)

86. It is believed that serial processors are a key to performing human tasks such as seeing, hearing, and even thinking. (T p442)

87. Massive parallelism refers to computers with hundreds, thousands, or even more processors; so many processors, in fact, that a different processor may be set to work on each different activity that may be accomplished in parallel. (T p442)

Fill-in-the-Blanks

1. **Artificial intelligence** is the study of making machines that exhibit intelligent behavior.

Robotics

2. Many **industrial** robots used in today's factories are controlled by small computers, but they could easily be built with simple electrical and mechanical parts.

The Mechanical Human

3. **Anthropomorphic** robots have the form and attributes of humans.

4. The electromechanical people and animals developed by Disney's "Imagineers" are called **audio-animatrons** .

A Brief History of Robotics

5. The word robot was coined by the Czech playwright **Karel Capek** .

6. The father of industrial robots is **Joseph Engelberger** .

7. **First** generation robots are little more than dumb mechanical manipulators.

8. **Second** generation robots have elementary vision and touch capabilities, and they may be equipped with arms, hands, and optic sensors.

9. **Third** generation robots will have highly developed sensory capabilities and limited intelligence.

10. Highly intelligent **fourth** generation robots will probably not appear until the late 1990s.

Productivity Concerns

11. The country that is the largest user of of industrial automation and robots in the world is **Japan**.

12. **Flexible** automation refers to the use of industrial robots in small batch production environments.

Enrichment Piece: Asimov's Laws of Robotics

13. The term robotics was coined by **Isaac Asimov**.

14. The father of fictional robotics is **Isaac Asimov**.

15. According to Isaac Asimov, a robot may not **injure** a human being.

Industrial Robots: "Steel Collar" Workers

16. **Industrial** robots are the mechanical arms used in industry for jobs such as welding, spray painting, lifting, and manipulating nuclear fuel rods.

17. **Industrial** robots available today cost a few dollars per hour to run, work around the clock, and are productive almost all the time.

How Robots Work

18. **Pick-and-place** robots are relatively simple dumb devices that repeatedly perform only one task sequence on command.

19. With **programmable** robots, a person leads the robot's arm manually through the series of movements and tasks it is to perform.

20. **Computerized** robots are software programmable.

21. **Sensory** robots have artificial vision or touch capabilities.

22. Once robot touch and vision capabilities are sufficiently developed, **assembly** robots will be put to work on assembly lines.

23. The **bin picking** problem must be solved before assembly robots will be truly effective.

Robot Vision

24. Robot **vision** systems currently under development convert signals from a TV camera into tens of thousands of dots of varying intensities.

An Artificial Sense of Touch

25. Robotic **skin** is being developed from thin sheets of rubber lined with a net of fine wires.

Household Robots

26. A typical **home** is a far more complex place for a robot than an assembly line.

Enrichment Piece: Clapp's Laws of Robotics

27. **Neale Clapp**, a social scientist and industrial relations consultant, has suggested guidelines for smoothing labor-management relations when installing robots.

What Will People Do?

28. The **Japanese** have robot factories run by robots.

Philosophical Issues

29. **Marvin Minsky** of MIT is one of the founding fathers in the field of artificial intelligence.

30. Genetic engineers are now tinkering with **DNA**, the building block of life.

The Turing Test of Intelligence

31. **Alan Turing** proposed a test for machine intelligence.

32. A machine is considered to have passed the **Turing** Test if the answers it gives the

interrogator cannot be distinguished from the answers given by a human.

Expert Systems

33. Computer systems that can provide intelligent answers to questions in specialized areas are called **expert systems** .

Expert System Components

34. The part of an expert system that contains the factual knowledge and the rules of thumb of a given area of interest is the **knowledge base**.

35. The part of an expert system that is a set of programs that reasons from the information contained in the knowledge base is the **inference engine** .

36. The **natural language processor** of an expert system is a set of programs that enables the user and computer to carry on a user-friendly dialogue.

37. The area of memory in which a user's problem statement is placed in an expert system, is the **workspace** .

38. The area of memory that the computer uses as a scratchpad in an expert system is called the **workspace** .

40. The **explanation system** of an expert system details for the user precisely how each conclusion is reached.

41. To obtain information for an expert system, a person called a **knowledge engineer** interviews a human expert in the field of interest.

42. The human expert that supplies information for an expert system is called a **domain** expert.

Natural Language Processing

43. **Natural language processing** is concerned with the problems of developing machines that understand languages such as English, French, German, and Spanish.

44. The central problem in the field of natural language processing is called **natural language translation** .

45. **Ambiguity** refers to the uncertainty of meaning, and is a key problem in the field of natural language processing.

Heuristics

46. A **heuristic** is a rule of good judgment or rule of thumb based on experience or intuition that human experts may use to attempt to solve a hard problem.

Enrichment Piece: Eliza and the Turing Test

47. The expert system, developed at MIT by Joseph Weizenbaum, that played the part of a psychiatrist is called **Eliza**.

The Game of Chess

48. Artificial intelligence researchers have viewed as a worthwhile goal the creation of a **chess** playing computer that could beat the world's best human player.

Lisp: The Programming Language of Artificial Intelligence

49. The programming language used most often in the field of artificial intelligence is **Lisp**.

50. The Lisp programming language is ideal for storing the kinds of **knowledge** structures like the if-then rules used in many expert systems.

51. Lisp programs run efficiently on very powerful special-purpose computers called **Lisp machines**.

Serial Versus Parallel Computers

52. Computers with a few processors that function simultaneously are called **multiprocessors**.

53. **Massive parallelism** refers to computers with hundreds, thousands, or even more processors.

54. A **Serial processor** processes one task at a time.

Chapter 17
Computers and Medicine

Multiple Choice

1. The low cost and small size of microprocessors has made it possible to build ______ into medical equipment.
   (a) intelligent control *
   (b) implantable devices
   (c) glucose
   (d) electrocardiograms
   (e) tomography

2. The medical expert system that is used to help diagnose diseases is called ______.
   (a) PET
   (b) CADUCEUS *
   (c) CAT
   (d) DSR
   (e) ELIZA

3. The CADUCEUS diagnosis system was developed by Jack Myers and ______.
   (a) Reginald Spencer
   (b) Godfrey Hounsfield
   (c) Harry Pople *
   (d) Allan Cormack
   (e) Richard Stengel

Computer-Based Scanning Systems

4. Devices that use techniques such as x-rays, gamma rays and ultrasound to obtain data from inside the human body are called ______.
   (a) intelligent control devices
   (b) pacemakers
   (c) catheters
   (d) computer-based scanning systems *
   (e) irradiation devices

5. The use of scanning systems to obtain information from inside the body without surgery is considered to be a(n) ______ diagnostic technique.
   (a) invasive
   (b) irradiation

(c) monitoring
(d) catheterization
(e) noninvasive *

The CAT Scan

6. The Nobel Prize for medicine was awarded to Allan Cormack and ______ for their work in developing the CAT scan.
(a) Harry Pople
(b) Godfrey Houndsfield *
(c) Jack Myers
(d) Kevin McKean
(e) Reginald Spencer

7. ______ means the description of slices.
(a) Tomography *
(b) Slicography
(c) Angiography
(d) Cathetography
(e) Electrocardiogram

8. A certificate of need is required by the ______ before a CAT scan can be purchased.
(a) AMA
(b) state government
(c) federal government *
(d) hospital
(e) SEOPF

The PET Scan

9. The scanning device used to monitor glucose activity is the ______ scan.
(a) DSR
(b) DNA
(c) CAT
(d) AMA
(e) PET *

10. The ______ scan is being used to detect and treat epilepsy.
(a) AMA
(b) DSR
(c) DNA
(d) PET *
(e) CAT

11. The computerized scanning method that is especially used to diagnosis heart and blood vessel problems is ______.
    (a) computer-assisted tomography
    (b) computer-axial tomography
    (c) angiography *
    (d) dynamic spatial reconstructor
    (e) catheterization

Dynamic Spatial Reconstructor

12. The ______ scan provides three-dimensional motion pictures of organs.
    (a) DSR *
    (b) CAT
    (c) DNA
    (d) PET
    (e) AMA

Heart Research

13. A(n) ______ is a device that helps control the functions of the heart.
    (a) electrocardiogram
    (b) pacemaker *
    (c) catheter
    (d) curare
    (e) eserine

14. ______ implantable devices are now under development that will allow doctors to alter them without the need for surgery.
    (a) Computer-controlled laser
    (b) Genetic engineering
    (c) Ultrasound
    (d) Electrocardiogram
    (e) Externally programmable control *

15. A(n) ______ is a graphical record of a patient's heart activity.
    (a) heart alert
    (b) angiograph
    (c) catheter
    (d) interferon
    (e) electrocardiogram *

16. The ______ is a system that sends emergency help to someone having a heart attack.
    (a) electrocardiogram
    (b) interferon

(c) catheter
(d) heart alert *
(e) angiography

Cancer

17. The precise control of ______ devices in cancer therapy is often performed by computer.
(a) tomography
(b) electrocardiogram
(c) irradiation *
(d) angiography
(e) computer-based scanning system

18. Scientists in the field of ______ are concerned with the structure of DNA.
(a) genetic engineering *
(b) noninvasive diagnosis
(c) drug conflicts
(d) angiography
(e) schizophrenia

19. The building block of life is ______.
(a) glucose
(b) DNA *
(c) interferon
(d) curare
(e) eserine

Organ Matching

20. The SEOPF is an example of a(n) ______ organization.
(a) cancer research
(b) heart research
(c) organ matching *
(d) medical library
(e) drug conflict warning

Anesthesiology

21. A goal of ______ is to reduce the amounts of drugs needed to make a patient unconscious and immobile.
(a) computerized patient monitoring
(b) computer-based scanning
(c) computerized medical libraries

(d) computer-controlled anesthesiology *
(e) drug conflict warnings

Surgery

22. With ______ doctors have been able to kill tumors and fuse detached retinas.
(a) computer-controlled anesthesiology
(b) computer-based scanning
(c) computerized patient monitoring
(d) intensive care units
(e) computer-controlled lasers *

Intensive Care

23. The most carefully staffed and equipped area in a hospital outside the operating room is probably the ______.
(a) ICU *
(b) DNA
(c) SEOPF
(d) DSR
(e) AMA

24. With computerized ______ a person's vital signs are closely tracked and intravenous feeding and drug flow are controlled.
(a) angiography
(b) patient monitoring *
(c) scanning
(d) lasers
(e) medical histories

Obtaining Medical Histories

25. Small computer systems are used in many doctor's offices to obtain ______ from patients via an interactive dialogue.
(a) drug warnings
(b) certificate of need
(c) electrocardiograms
(d) medical histories *
(e) organ matching

26. A complete medical history can be placed on a ______ that can be updated.
(a) pacemaker
(b) implantable device

(c) smart card *
(d) heart alert
(e) electrocardiogram

Drug Conflict Warnings

27. Computers can help doctors and pharmacists by issuing ______.
(a) heart alerts
(b) electrocardiograms
(c) medical libraries
(d) organ matching
(e) drug conflict warnings *

Searching Medical Libraries

28. Many doctors use ______ to locate important new developments in the medical field quickly.
(a) drug conflict warnings
(b) computerized medical libraries *
(c) computerized organ matching
(d) smart cards
(e) heart alert

Artificial Organs and Body Parts

29. There is research being done with microprocessor-controlled ______ such as hearts, lungs and livers.
(a) artificial organs *
(b) pacemakers
(c) lasers
(d) electrocardiograms
(e) organ matching

True/False

Introduction

1. Using a microprocessor to control a medical device provides the device with what is called intelligent control. (T p450)

2. With computerized diagnostic tools, exploratory surgery will be greatly increased. (F p450)

Computer Doctors

3. CADUCEUS is a computerized medical diagnosis system capable of diagnosing over 500 diseases. (T p451)

4. One problem with CADUCEUS is that it tends to order more lab tests than human doctors would normally order. (F p451)

5. CADUCEUS tends to get confused when a patient has several unrelated ailments. (T p451)

Computer-Based Scanning Systems

6. One advantage of the computerized scanning techniques such as CAT, PET, and untrasound is that they are noninvasive. (T p452)

The CAT Scan

7. The most significant example of computers in modern medicine is probably the PET scan. (F p452)

8. The CAT scan produces pictures of "slices" of the human body. (T p452)

9. The 1979 Nobel Prize for medicine or physiology was awarded to Godfrey Hounsfield and Allan Cormack for the development of the CAT scan. (T p452)

10. The CAT scan is one of the most economical medical devices in use today. (F p452)

11. Tomography means "the description of slices," an accurate explanation of what the CAT scan provides. (T p452)

12. CAT scan pictures of a patient could be taken by conventional photography but only at much higher cost. (F p454)

Elscint's Newest Marvel

13. Elscint's gamma camera can actually display moving pictures of internal organs. (T p453)

14. Elscint's gamma camera requires so much computer processing of information from the patient's body that the displays are delayed approximately 30 minutes. (F p453)

The PET Scan

15. The PET scan can help map brain activity. (T p454)

16. The PET scan monitors glucose (sugar) activity to determine areas of abnormal metabolism (the rate of "body function"). (T p455)

17. The CAT scan is especially useful for determining if a patient is manic-depressive or schizophrenic. (F p455)

18. The PET scan is useful in the treatment of epilepsy. (T p455)

19. The PET scan can eliminate the need for patients to undergo painful and dangerous catheterization. (T p455)

Dynamic Spatial Reconstructor

20. The state of the art in computerized x-ray scanning technology is the Dynamic Spatial Reconstructor. (T p456)

21. The DSR can look through the muscle tissue of an organ to show only the blood vessels on the display screen. (T p456)

Heart Research

22. With today's heart pacemakers, "It's like putting a computer in your chest." (T p457)

23. One problem common to implantable computerized control devices like heart pacemakers is that they must be surgically removed to be reprogrammed. (F p456)

Cancer

24. Cancer is one disease for which computers are simply of no help whatsoever. (F p457)

25. Genetic engineers use computers to study the structure of DNA, said to be the building block of life. (T p458)

Organ Matching

26. Computers facilitate organ transplants--particularly in organ matching. (T p458)

27. SEOPF is an organization concerned with using computers to control the irradiation of cancer patients. (F p458)

28. Because of the fact that organs must be transplanted soon after the donor's death, it is unlikely that organ matching systems will match donors and recipients throughout the world. (F p458)

Anesthesiology

29. One of the primary research goals in computer-assisted anesthesiology is to reduce the amount of anesthetics that need to be administered to make a patient unconscious and immobile. (T p459)

Intensive Care

30. The intensive care unit in any hospital is probably the most carefully staffed and well-equipped area in the hospital. (F p459)

31. Computerized patient monitoring eliminates the need for highly skilled personnel in intensive care units. (F p459)

Obtaining Medical Histories

32. Patients are generally reluctant to supply their medical histories to interactive computer systems, even though they know this can help reduce the cost of medical care. (F p460)

33. One fascinating phenomenon has been that patients seem more willing to "doscuss" certain personal problems with a computer than with a nurse or doctor. (T p460)

34. Today, smart cards are available that can contain a patient's complete medical history in a tiny yet high-capacity computer memory which is updated every time the person receives medical attention. (T p460)

Drug Conflict Warnings

35. Pharmacists and physicians automatically check all prescriptions they give a patient with computerized database systems to make sure that there are no dangerous conflicts. (F p460)

36. Because drug conflict warning systems save lives, there is little concern about the fact that these systems may invade an individual's privacy. (F p460)

Searching Medical Libraries

37. Today, virtually any facts doctors need to know are instantaneously available at their fingertips with the use of computerized medical libraries. (F p460)

Artificial Organs and Body Parts

38. Computers today can produce customized, tissue-matched artificial organs on demand. (F p461)

Fill-in-the-Blanks

1. The low cost and small size of microprocessors has made it possible to build **intelligent** control into medical diagnostic, monitoring, and control equipment.

Enrichment Piece: Computer Doctors

2. **CADUCEUS** is a medical expert system used to help doctors diagnose diseases.

3. Jack Myers and Harry Pople developed the medical expert system called **CADUCEUS** .

Computer-Based Scanning Systems

4. **Computer-based scanning** systems use computers and techniques such as x-rays, gamma rays, and ultrasound to obtain information about a person's internal organs.

5. The various scanning techniques are considered to be **noninvasive** diagnostic techniques because they do not use surgery in order to obtain a precise picture of the inside of the human body.

The CAT Scan

6. Godfrey Hounsfield and Allan Cormack shared a Nobel Prize for their development of the **CAT** scan.

7. The CAT scan uses the principles of **tomography** , which means the "description of slices."

8. Because of the high cost of CAT scans, the federal government requires hospitals to file a **certificate of need** before being allowed to purchase a machine.

The PET Scan

9. The **PET** scan helps to map brain activity by monitoring glucose activity to determine areas of abnormal metabolism.

10. The **PET** scan is useful in the treatment of epilepsy.

11. A form of computerized scanning being used to diagnosing heart and blood vessel problems is called **angiography** .

Dynamic Spatial Reconstructor

12. The **Dynamic Spatial Reconstructor** (DSR) provides three-dimensional motion pictures of organs.

Heart Research

13. An implantable device used to monitor heart functions is the **pacemaker** .

14. Researchers are building **externally programmable** control devices that perform many more functions than the common pacemaker and can be altered without surgery to react to changes in the patient's heart activity.

15. **Electrogcardiograms (EKGs)** are graphical records of a patient's heart activity.

Cancer

16. Computers are used for the precise control of irradiation devices used in **cancer** therapy.

17. Scientists in the field of **genetic** engineering are concerned with the structure of DNA.

Organ Matching

18. Computers are used to facilitate organ transplants; they are particularly useful in organ **matching** .

19. Organizations such as **SEOPF** in Richmond, Virginia, maintain computerized databases of registered organ recipient candidates.

Anesthesiology

20. One of the primary goals in computer-assisted **anesthesiology** is to reduce the amount of anesthetics that need to be administered to make a patient unconscious and immoble.

Surgery

21. In delicate eye surgery, computer-controlled **lasers** have been used to kill tumors and to fuse detached retinas.

Intensive Care

22. The **intensive care unit (ICU)** in any hospital is probably the most carefully staffed and well equipped area outside of the operating room itself.

23. Computerized patient **monitoring** is being used successfully for patients in intensive care to keep track of how they're progressing.

Obtaining Medical Histories

24. Computers are being used in hospitals and in doctors offices for the purpose of obtaining **medical** histories from patients.

25. **Smart** cards are now available that can contain a complete medical hsitory of a patient, and can be updated by a computer device every time the patient receives medical attention.

Drug Conflict Warnings

26. Computers are being used to catalogue facts about known drug interactions and to provide pharmacists and doctors with **drug conflict** warnings.

Searching Medical Libraries

27. Vast amounts of medical information are being kept in computerized **medical libraries** to help medical professionals locate information quickly.

Artificial Organs and Body Parts

28. Many researchers feel that microprocessor-controlled **artificial** organs, such as hearts, lungs, kidneys, and livers, may soon be in wide use.

Chapter 18
Computers and the Handicapped

Multiple Choice

1. ______ capabilities are those senses and movements of a handicapped person that still function properly.
   (a) Rehabilition
   (b) Residual *
   (c) Quadriplegic
   (d) Biomedical
   (e) Redundant

Rehabilitation Engineering

2. Applying technology to improving the lives of the handicapped is called ______.
   (a) biomedical engineering
   (b) information amplification
   (c) rehabilitation engineering *
   (d) computer-assisted education
   (e) residual capabilities

3. Both handicapped individuals and members of their families have been able to contribute to the field of ______ by developing their own applications for computers.
   (a) rehabilitation engineering *
   (b) residual capabilities
   (c) computer-assisted education
   (d) biomedical engineering
   (e) information amplification

Information Amplification

4. Through ______, computers can help the handicapped to work at more normal speeds.
   (a) EDP
   (b) database integrity
   (c) redundancy
   (d) information amplification *
   (e)device independence

Computers and Blind People

5. ______ uses raised dots embossed on paper to aid the blind in reading.
   (a) Computer-assisted education
   (b) Optacon
   (c) Minspeak
   (d) Homework
   (e) Braille *

The Orator

6. The ______ device allows a computer to speak its output.
   (a) Minspeak
   (b) Braille computer terminal
   (c) Orator *
   (d) Optacon
   (e) Speech recognizer

7. The ______ method of speech uses prerecorded human speech for playback.
   (a) synthetic speech
   (b) stored speech *
   (c) speech recognition
   (d) Minspeak
   (e) Deafnet

8. The ______ method of speech uses speech formation rules that are stored in the computer.
   (a) Deafnet
   (b) Minspeak
   (c) stored speech
   (d) synthetic speech *
   (e) speech recognition

9. The greatest advantage of ______ is its almost unlimited vocabulary.
   (a) synthetic speech *
   (b) speech recognition
   (c) Minspeak
   (d) Deafnet
   (e) stored speech

10. With ______ the quality of the speech produced is very good.
   (a) speech recognition
   (b) Deafnet
   (c) synthetic speech

(d) Minspeak
(e) stored speech *

11. With the ______ computer device, words and and phrases can be spoken as they are typed.
(a) Optacan
(b) Deafnet
(c) Orator *
(d) Minspeak
(e) Plato

Aids for Partially Sighted People

12. The ______ is a device that enables people who are partially sighted to become familiar with using a computer.
(a) Optacon
(b) Orator
(c) Large Print Video Treminal *
(d) softcopy Braille output terminal
(e) Abilityphone

Braille Computer Terminals

13. With the use of ______ terminals blind people are able to hold computer-related jobs.
(a) Homework
(b) Braille computer *
(c) Minspeak
(d) Optacon
(e) PLATO

14. Microprocessors cause a set of metal pins to move up and down on a flat surface with the ______ device.
(a) Orator
(b) Optacon
(c) PLATO
(d) softcopy Braille output *
(e) Large Print Video

15. The ______ device is like a CRT terminal to a sighted person.
(a) Large Print Video
(b) Orator
(c) PLATO
(d) Optacon
(e) softcopy Braille output *

Sonar "Sight"

16. One device that provides sight uses the ______ focusing element of the Polaroid cameras.
    (a) sonar *
    (b) Braille
    (c) optical
    (d) eye
    (e) lens

A Television "Eye"

17. A system being developed for the blind uses a ______ that is mounted on the person's shoulder; it sends signals to a microcomputer that informs the person of nearby obstacles.
    (a) Polaroid camera
    (b) sonar element
    (c) television camera *
    (d) telesensory system
    (e) lens

Reading by Vibrations

18. ______ is a device that scans letters and converts them into vibrations that a person feels by fingertip.
    (a) Orator
    (b) Large Print Video terminal
    (c) Kurzweil Reading machine
    (d) Optacon *
    (e) Braille computer terminal

The Kurzweil Reading Machine

19. The ______ uses OCR technology.
    (a) Optacon
    (b) Kurzweil Reading Machine *
    (c) Orator
    (d) Braille computer terminal
    (e) Large Print Video terminal

20. The Kurzweil Reading Machine speaks words through a ______.
    (a) speech recognizer
    (b) stored speech program
    (c) synthetic speech program

(d) text-to-speech translator
(e) speech synthesizer *

21. Of the following, ______ is not a function of the operator keys of the Kurzweil Reading Machine.
(a) skim through text
(b) mark text to return at a later time
(c) turn the page *
(d) repeat a single line or word
(e) resume operation

Computers and Deaf People

22. ______ is the American sign language that is used by both deaf and nondeaf to communicate.
(a) PLATO
(b) Ameslan *
(c) Minspeak
(d) Deafnet
(e) Braille

23. Ameslan can be used over telephone lines with a microprocessor device developed by ______ experts.
(a) computer graphics *
(b) computer science
(c) biomedical
(d) rehabilitation
(e) communication

Electronic Mail for the Deaf

24. ______ is an electronic mail service for the deaf.
(a) PLATO
(b) Homework
(c) GTE Telenet
(d) Deafnet *
(e) TIC

Speech Recognition Devices for the Deaf

25. The ability of a computer to understand spoken words is called ______.
(a) speech synthesis
(b) stored speech
(c) synthetic speech

(d) speech synthesizer
(e) speech recognition *

Computers and Deaf-Blind People

26. Deaf-blind people can communicate through ______.
(a) eye-tracking
(b) vibrations *
(c) Ameslan
(d) TIC
(e) speech synthesis

Computers and Nonvocal Disabled People

27. The ______ displays letters on a screen, the user can selects letters to help form words.
(a) Tufts Interactive Communicator *
(b) Optacon
(c) Kurzweil Machine
(d) Minspeak
(e) Orator

28. With the ______ device, the letters of the alphabet are displayed in order of their frequency of use in the English language.
(a) Orator
(b) Minspeak
(c) Tufts Interactive Communicator *
(d) Optacon
(e) Kurzweil Reading

29. ______ devices determine what a handicapped person is looking at.
(a) Speech recognition
(b) Text-to-speech translator
(c) Optacon
(d) Sonar
(e) Eye-tracking *

The Abilityphone

30. The ______ is a computerized telephone device that can call for help in emergencies, and can perform a wide variety of other functions to assist handicapped people.
(a) Tufts Interactive Communicator
(b) Minspeak
(c) Orator

(d) Abilityphone *
(e) Optacon

31. The ______ uses both a speech synthesizer and a display screen to aid its users.
(a) Optacon
(b) Abilityphone *
(c) Tufts Interactive Communicator
(d) Ameslan
(e) Minspeak

32. Of the following ______ is not a feature of the Abilityphone.
(a) speech-to-text translator *
(b) environmental controller
(c) alarm clock
(d) calendar
(e) calculator

Minspeak

33. ______ uses a keyboard that contains pictures to represent concepts.
(a) Tufts Interactive Communicator
(b) Optacon
(c) Abilityphone
(d) Minspeak *
(e) Orator

34. The ______ system analyzes concepts entered by the user and selects an idea that best conveys the idea of the user.
(a) Orator
(b) Tufts Interactive Communicator
(c) Optacon
(d) Abilityphone
(e) Minspeak *

Computers and the Paralyzed

35. A quadriplegic's major problem is one of ______.
(a) communication
(b) sight
(c) mobility *
(d) hearing
(e) speech

36. Robotic devices are being developed with ______ devices to help quadriplegics.

(a) speech synthesis
(b) voice-recognition *
(c) eye-tracking
(d) Braille
(e) sonar

Restoring Movement to Paralyzed Limbs

37. ______ is known for his work in restoring movement to paralyzed limbs through microprocessor implantation and computer control.
(a) Jerrold Petrofsky *
(b) Peter Duran
(c) Louis Braille
(d) Raymond Kurzweil
(e) John Stapp

The Handicapped at Work

38. ______ is a computer-assisted education system developed by CDC.
(a) HOMEWORK
(b) TIC
(c) Braille
(d) Deafnet
(e) PLATO *

Computers and Special Education

39. Which of the following does not pertain to the use of computers to help educate people with learning disabilities? ______
(a) self-paced presentation
(b) customized content
(c) infinite patience
(d) fun to use
(e) eye-tracking *

40. ______ are used to help restore hand-eye coordination in stroke victims.
(a) Work-at-home programs
(b) Eye-tracking devices
(c) Video games *
(d) Speech recognition devices
(e) Braille computer terminals

Biomedical Engineering

41. The field of ______ is actively developing replacement body parts.
    (a) rehabilitation engineering
    (b) genetic engineering
    (c) computer-assisted education
    (d) biomedical engineering *
    (e) information amplification

42. Computers analyze body movements to help doctors and engineers design ______.
    (a) artificial body parts *
    (b) speech synthesizers
    (c) speech recognition devices
    (d) video games
    (e) bionics

True/False

Introduction

1. A computerized "voice board" enables a nonvocal person to speak; the person touches words or phrases and the device then speaks through a speech recognizer. (F p466)

2. Residual capabilities are those senses provided to handicapped people through computerized devices. (F p466)

Rehabilitation Engineering

3. Applying technology to improving the lives of handicapped people is called rehabilitation engineering. (T p466)

Information Amplification

4. A handicapped person capable of typing, but only at a slow speed, could use abbreviations that a computer program would then expand to full words and phrases; this is an example of information amplification. (T p467)

Computers and Blind People

5. Only about 10% of the books published in the United States are available in Braille. (F p467)

6. Blindness, in varing degrees, affects about 2 million Americans. (T p467)

7. The Braille cell is three dots high and two dots wide. (T p468)

8. With the number of dots available in the Braille cell, it is possible to form 256 basic characters or sounds. (F p468)

9. The speech sounds in Braille are called digraphs. (T p468)

The Orator

10. The Orator is a device used primarily by nonvocal people to provide them with a pleasant sounding artificial voice. (F p469)

11. The Orator uses synthetic speech rather than stored speech. (T p469)

12. In the synthetic speech method, human speech is recorded and stored in the computer's memory for playback. (F p469)

13. In general, stored speech is of a higher quality than synthetic speech, but synthetic speech systems are improving rapidly. (T p469)

14. In general the vocabulary of stored speech systems is larger than that of synthetic speech systems. (F p469)

15. One problem that the designers of text-to-speech translators face is that the same thing can often be said in several ways. (T p469)

16. The Orator can spell out words it doesn't understand. (T p469)

Aids for Partially Sighted People

17. One of the most common types of computerized devices for partially sighted people is the Large Print Video Terminal. (T p469)

Braille Computer Terminals

18. The most popular reading aid for the blind today is the Kurzweil Reading Machine. (F p470)

19. The raised dot printing method called Braille was developed in 1824 by Louis Braille who was himself blind. (T p470)

20. All Braille output devices are hard copy devices; they produce raised dots on paper. (F p470)

Sonar "Sight"

21. Only about 1% of the blind people in the United States use guide dogs. (T p470)

22. The most popular mobility aid for blind people is currently the "sonar sight" system. (F p470)

23. It is interesting to note that the sonar focussing element made popular in Polariod cameras has been used to develop a mobility aid for the blind. (T p470)

A Television Camera "Eye"

24. Researchers at the Smith-Kettlewell Institute specialize in developing aids for the deaf. (F p470)

25. One mobility aid for the blind uses a television camera, computer, and speech synthesizer to describe to the blind person the nature of, and distance to, nearby obstacles. (T p470)

Reading by Vibrations

26. With the Optacon, a blind person may "read by vibrations." (T p471)

27. With the Optacon a person can read text at normal reading and conversation rates. (F p471)

The Kurzweil Reading Machine

28. Perhaps the most exciting application of computer technology today for the blind is the Kurzweil Reading Machine. (T p471)

29. The Kurzweil Reading Machine speaks at about normal conversation speed. (T p471)

30. The Kurzweil Reading Machine uses the stored speech method of speech generation. (F p471)

31. The Kurzweil Reading Machine stores more than 1000 English grammer rules as well as 15000 exceptions to assist in the text-to-speech translation process. (F p471)

32. The first individual to purchase a Kurzweil reading Machine for personal use was rock star

Ray Charles. (F p471)

Computers and Deaf People

33. Today, people gather information much more by reading than by hearing. (F p471)

34. In the United States, more than 13 million people have significant hearing loss. (T p472)

35. Approximately 2 million Americans are totally deaf. (T p472)

36. TDD capabilities are primarily useful for blind people. (F p472)

37. Ameslan is a means of communicating with deaf people. (T p472)

38. Deaf people may not communicate over the telephone. (F p472)

39. Deaf people may actually "speak" Ameslan over the telephone. (T p473)

40. One problem with trying to "speak" Ameslan over the telephone is that the bandwidth of a telephone is simply too narrow to transmit a full video picture. (T p473)

Electronic Mail for the Deaf

41. The Deafnet system operated by GTE Telenet is a 24 hour per day electronic mail system for the deaf. (T p473)

42. One popular feature of Deafnet is its computerized electronic bulletin board system that provides much useful information for the deaf. (T p473)

Speech Recognition Devices for the Deaf

43. Other than complete restoration of hearing, perhaps the ultimate breakthrough for the hearing impaired will come when speech recognition devices enable computers to recognize human speech easily. (T p474)

Computers and Deaf-Blind People

44. More than 10,000 people in the United States are both deaf and blind. (T p474)

45. Deaf-blind people can receive communications by Braille and by computer-controlled devices that signal by vibrations. (T p474)

Computers and Nonvocal Disabled People

46. The Tufts interactive Communicator helps a nonvocal person who can't use a keyboard form words one letter at a time by operating a switch. (T p474)

47. With the Tufts Interactive Communicator, the letters of the alphabet are shown on a screen in alphabetical order for the convenience of the user. (F p474)

48. A touch-sensitive pad with common words and phrases can help even a nonvocal quadriplegic person communicate. (T p475)

49. For people with severe loss of muscle control, computerized eye-tracking devices have been developed that determine what the eye is looking at and thus what the disabled person wants to say. (T p475)

The Abilityphone

50. The Abilityphone is primarily a telephone aid for deaf people. (F p476)

Minspeak

51. The Minspeak system uses a conventional typewriter keyboard and converts words to synthetic speech. (F p476)

52. Minspeak is an example of an information amplification system. (T p476)

53. The Minspeak system prestores thousands of sentences. (T p476)

54. Researchers have found that with the Minspeak system a relatively small number of keys and keystrokes enable the user to express a huge number of common thoughts. (T p476)

Computers and the Paralyzed

55. A quadripplegic is a person handicapped by the loss of movement in both arms and both legs. (T p476)

56. Quadriplegics often retain the ability to make shoulder, neck, and head movements, and their senses of taste, sight, hearing, and smell function properly. (T p476)

57. A quadriplegic's main problem is usually communication. (F p476)

58. A robotic arm with a voice recognizer can be an effective aid for a quadriplegic. (T p476)

Restoring Movement to Paralyzed Limbs

59. In paralysis, the problem is often that damaged nerves prevent control signals from the brain from reaching otherwise healthy muscles. (T p478)

60. Professor Jerrold Petrofsky has been working on techniques for using computers to repair damaged nerves to help paralyzed people walk. (F p478)

61. Professor Jerrold Petrofsky uses implanted microprocessors to generate simulated brain signals to help paralyzed limbs move. (T p478)

62. It is entirely possible that computers may eventually help restore mobility even to quadriplegics. (T p478)

The Handicapped at Work

63. One widely touted work-at-home program is HOMEWORK developed by Control Data Corporation in 1978. (T p478)

64. The Audio Typing Unit developed by IBM helps blind people be employed as office typists. (T p480)

Computers and Special Education

65. Computers with color graphics, sound, and interaction capabilities can help hold an individual's attention; this is espically important because many handicaps affect a person's concentration and motivation as well. (T p479)

66. Video games can be rehabilitation tools. (T p480)

Biomedical Engineering

67. One goal of the field of biomedical engineering is the development of bionic body parts. (T p481)

68. Microprocessors have been built into implantable insulin pumps that regulate the level of blood sugar in disbetics. (T p481)

69. The MIT Knee has a built-in microprocessor that controls the complex knee movements. (T p481)

70. Head-controlled electronic wheelchairs have been developed for quadriplegics. (T p482)

71. Biomedical engineers use computers to analyze the movements of normal limbs; this helps them design artificial limbs. (T p483)

Fill-in-the-Blanks

1. The senses and movements that still function for handcapped people are referred to as **residual capabilities** .

Rehabilitation Engineering

2. Applying technology to improving the lives of the handicapped is called **rehabilitation engineering** .

Information Amplification

3. Through information **amplification** , computers can help handicapped people work at more normal speeds.

Computers and Blind People

4. The **Braille** language is notations of raised dots embossed in paper that allows people to read by touch.

The Orator

5. A device that uses speech synthesis to enable a computer to speak its outputs is called the **Orator** .

6. With **stored speech** , human speech is recorded and stored in the computer's memory for playback.

7. The **synthethic speech** method of speech generation, stores several speech formation rules that help the computer analyze standard English pronunciations.

Aids for Partially Sighted People

8. The **Large Print Video Terminal** is a computer terminal for the partially sighted.

Braille Computer Treminals

9. The Braille language was developed by **Louis Braille**.

10. **Braille** computer terminals enable blind people to hold productive computer-related jobs.

11. A microprocessor causes a set of metal pins to move up and down on a flat surface to form a line of Braille characters in **soft copy** Braille output devices.

12. Output devices for the blind that are like what CRT screens are to sighted people are called **soft copy Braille output** devices.

Sonar "Sight"

13. A device that was developed for the blind provides sonar "sight" by using the sonar focusing element made popular in **Polaroid** cameras.

Reading by Vibrations

14. The **Optacon** is a device that enables a blind person to "read by vibrations."

The Kurzweil Reading Machine

15. The **Kurzweil Reading Machine** scans ordinary printed material; it recognizes individual symbols and letters, assembles them into words, and speaks the words via a speech synthesizer.

16. The Kurzweil Reading Machine speaks its output through a **speech synthesizer**.

Computers and Deaf People

17. **Ameslan** is the American sign language used by the deaf.

Electronic Mail for the Deaf

18. The electronic mail system operated by GTE Telenet for the deaf is called **Deafnet**.

Speech Recognition Devices for the Deaf

19. Perhaps the ultimate breakthrough for the hearing impaired will come when **speech recognition** technology improves to the point that computers can easily recognize human speech.

Computers and Nonvocal Disabled People

20. The **Tufts Interactive Communicator** is a device that helps a person form words a letter at a time by flipping a switch.

21. People with severe loss of muscle control, **eye-tracking** devices have been developed that determine what the eye is looking at.

The Abilityphone

22. People with disabilities can receive help in an emergency from the **Abilityphone** device.

23. The emergency device that can call for help, reminds people to take medication, answers the phone, and periodically asks "Are you OK?" is the **Abilityphone** .

Minspeak

24. The **Minspeak** system uses a unique keyboard with keys that contain pictures to represent concepts a handicapped person wishes to convey.

Computers and the Paralyzed

Restoring Movement to Paralyzed Limbs

25. **Jerrold Petrofsky** , a professor at Wright State University, is applying computer technology to helping restore movement to paralyzed limbs.

The Handicapped at Work

26. Many companies offer **work-at-home** programs to handicapped people that allow them to hold productive jobs without the need of traveling to a job site.

27. A work-at-home program develped by Control Data Corporation is called **HOMEWORK** .

Computers and Special Education

28. Computerized **video games** have been used by doctors to help improve hand-eye coordination, reaction time, memory, span of concentration, and problem solving skills for stroke and cerebral palsy victims.

Biomedical Engineering

29. The field of **biomedical engineering** is developing replacement body parts such as hearts, eyes, knees, and skin.

30. The popular TV program "The Six Million Dollar Man" featured a character that had **bionic** body parts.

31. The movements of of the human body are analyzed by computers to help doctors and engineers to design and construct **artificial body** parts that are customized to an individual's specific needs.

Chapter 19
Computers and Transportation

Multiple Choice

Computers and Automobiles

1. ______ has become one the world's largest computer manufacturers just to service its own needs.
   (a) General Motors *
   (b) Lockheed Corp.
   (c) Yamaha
   (d) Rand Corp.
   (e) Westinghouse

2. The system available in General Motors cars that monitors many engine functions is ______.
   (a) modulated displacement system
   (b) computer command control *
   (c) database/data communication system
   (d) fuel planning system
   (e) collision avoidance system

3. With the ______ system a microprocessor automatically reduces the number of cylinders in operation.
   (a) fuel planning
   (b) collision avoidance
   (c) modulated displacement *
   (d) database/data communications
   (e) computer command control

4. Of the following ______ is not a function of the computer for the Yamaha motorcycle.
   (a) status of the brakes
   (b) battery fluid
   (c) kickstand position
   (d) lock *
   (e) engine oil

5. Which of the following is not one of the prestored phrases available in automobiles? ______
   (a) Turn off your lights
   (b) Lock the doors
   (c) Take the key

(d) Fasten your seatbelt
(e) Turn on your lights *

6. In the St. Gotthard Tunne, the computerized control system does not monitor ______.
(a) traffic flow
(b) ice conditions *
(c) air quality
(d) lighting
(e) ventilation equipment

Car Pooling

7. ______ is a way to conserve fuel and reduce commuting costs.
(a) Car pooling *
(b) Traffic control
(c) Modulated displacement system
(d) Computer command control
(e) Gridlock

8. Of the following, which is not a consideration of a car pooling system? ______
(a) starting points
(b) passenger capacity of automobile
(c) reserving parking places *
(d) destinations
(e) working hours

Traffic Control

9. The synchronizing of traffic lights is a feature of ______.
(a) modulated displacement systems
(b) gridlock
(c) computer command control
(d) car pooling
(e) computerized traffic control *

10. Sharing cable networks helps reduce the cost of ______.
(a) gridlock
(b) car pooling
(c) computer command control
(d) computerized traffic control *
(e) modulated displacement systems

11. When intersections are blocked and become impassable ______ occurs.
(a) gridlock *

(b) modulated displacement
(c) traffic control
(d) car pooling
(e) computer command control

Emergency Service

12. Of the following items which is not supplied to an emergency road service dispatcher by a driver? ______
(a) problem with the car
(b) completion time *
(c) name
(d) location
(e) phone number

Driving Simulators

13. ______ are used in driver training programs to give students a wide range of driving situations and conditions.
(a) Electronic maps
(b) Car pools
(c) Collision avoidance sysyems
(d) Driving simulators *
(e) Traffic control computers

Computers and Air Travel

Air Traffic Control

14. ______ is the agency responsible for monitoring and controlling the air traffic over American skies.
(a) ICTS
(b) ATARS
(c) FAA *
(d) BCAS
(d) DABS

15. With the ______ system of air traffic control, ground-based radar and computer systems will keep track of all aircraft.
(a) FAA
(b) BCAS
(c) FCC
(d) ICTS

(e) DABS *

16. The device on an aircraft that automatically sends a reply when it receives a signal is called a ______.
(a) simulator
(b)transducer
(c) transponder *
(d) transformer
(e) DB/DC

17. ______ will work along with the DABS system by keeping track of plane positions, monitoring flight paths, and detecting potential collisions.
(a) ATARS *
(b) BCAS
(c) ICTS
(d) DB/DC
(e) TDA

18. The collision avoidance system that is air-based is the ______.
(a) DABS
(b) BCAS *
(c) ATARS
(d) TDA
(e) ICTS

19. The air traffic control and avoidance system that continues to operate even if ground-based systems fail is ______.
(a) ICTS
(b) TDA
(c) BCAS *
(d) DABS
(e) ATARS

The Holding Pattern

20. Planes circle airports in ______ waiting to get landing clearance.
(a) transponders
(b) voting schemes
(c) gridlocks
(d) holding patterns *
(e) fuel planning systems

21. In the following list which is not a consideration of computerized systems that reduce holding patterns? ______
(a) range

(b) time
(c) course
(d) altitude
(e) weather *

The Airborne Electronic Terrain Map System

22. The system that will allow pilots to fly safely at night, in poor weather conditions, or over unfamiliar territory is the ______.
 (a) Discrete Address Beacon System
 (b) Airborne Electronic Terrain Map System *
 (c) Beacon Collision Avoidance System
 (d) Automatic Traffic Advisory and Resolution Service
 (e) Intermediate Capacity Transit System

23. The system currently under development by the Air Force is the ______.
 (a) Airborne Electronic Terrain Map System *
 (b) Intermediate Capacity Transit System
 (c) Discrete Address Beacon System
 (d) Automatic Traffic Advisory and Resolution Service
 (e) Beacon Collision Avoidance System

24. Which of the following is not supplied by a navigational computer to a mapping computer? ______
 (a) latitude
 (b) longitude
 (c) weather *
 (d) course
 (e) altitude

Flight Simulators

25. With the use of ______ students can learn to fly different types of planes without leaving the ground.
 (a) holding patterns
 (b) transponders
 (c) fuel planning systems
 (d) collision aviodance systems
 (e) flight simulators *

Fuel Planning

26. Which of the following facts is not important in a computerized fuel planning system?

______
(a) weather conditions
(b) number of passengers
(c) destination
(d) latitude *
(e) cargo weight

Computerized Reservations Systems

27. With computerized ______ systems airlines can keep track of how many seats are available on a given flight.
(a) fuel planning
(b) electronic map
(c) reservation and ticketing *
(d) flight simulator
(e) ferrying

28. Reservations computer systems are examples of ______ systems.
(a) LIMs
(b) DB/DC *
(c) DABS
(d) BD/CD
(e) ATARS

Computers and Trains

Canada's Intermediate Capacity Transit System

29. The ______ is a heavily computerized transportation system in use in Canada.
(a) ICTS *
(b) LIM
(c) TDA
(d) DB/DC
(e) BCAS

30. ______ motors that work on the principle that opposite magnetic poles attract.
(a) Maglev
(b) ICTS
(c) TDA
(d) DB/DC
(e) LIM *

31. Trains that are pulled along by magnetic attraction use ______.
(a) TDA

(b) Maglev
(c) LIMs *
(d) DB/DC
(e) ICTS

32. In situations where high reliability is needed, computers may operate with a ______.
(a) tracking system
(b) voting scheme *
(c) database/data communications system
(d) collision avoidance system
(e) traffic control system

The Westinghouse People Mover System

33. The ______ operates at the Atlanta Airport, moving passengers between the main terminal and the concourses.
(a) People Mover System *
(b) Intermediate Capacity Transit System
(c) Planetran
(d) Train Dynamics Analyzer
(e) Magnetic Levitation Train

Magnetic Levitation Trains

34. ______ is accomplished by the principle that like poles repel each other.
(a) Linear induction
(b)Magnetic bifurcation
(c) TDA
(d) Magnetic levitation *
(e) ICTS

35. The ultimate train of the future may well be the ______.
(a) LIM
(b) ICTS
(c) People Mover
(d) TDA
(e) Planetran *

36. Trains that operate with maglev are in use in ______.
(a) Canada
(b) Germany
(c) Japan *
(d) United States
(e) England

"Tracking" Railroad Cars

37. With the use of ______ systems, railroad companies can determine the location, contents and weight of railroad cars.
    (a) voting
    (b) tracking *
    (c) database/data communications
    (d) fuel planning
    (e) reservation and ticketing

Train Simulators

38. Conrail uses the ______ simulator to train new engineers and sharpen the train-handling skills of experienced personnel.
    (a) ATARS
    (b) LIM
    (c) ICTS
    (d) TDA *
    (e) DABS

Computers and Ships

39. Which of the following is not a job of the computer in a computerized sailing ship? ______
    (a) charts electronic maps *
    (b) monitors wind speed and direction
    (c) operates the sails
    (d) controls the motors
    (e) monitors weather conditions

40. Using computerized ______, crew members can practice docking in ports around the world without even being on a ship.
    (a) sailing ships
    (b) collision avoidance systems
    (c) simulators *
    (d) traffic control
    (e) fuel planning

True/Flase

Computers and Automobiles

1. In 1984 fully equipped U.S.-built cars used 10 to 15 chips, compared with less than one chip per car in 1978. (T p490)

2. GM has actually become one of the world's largest chip manufacturers, just to service its own needs. (T p490)

3. With Computer Command Control, GM cars can actually repair themselves. (F p490)

4. With Computer Command Control, the number of cylinders in operation in an automobile engine is reduced for fuel efficiency as a car achieves crusing speed. (F p491)

5. Yahama announced the world's first computerized motorcycle. (T p491)

6. Today's speaking automobiles generally use synthetic speech generation to warn the driver to turn off the lights, lock the doors, and so on. (F p492)

Car Pooling

7. Car pooling systems are examples of database applications. (T p492)

Traffic Control

8. New York City has converted thousands of traffic lights to computer operation, and as a result motorists are already saving 20 million gallons of fuel per year. (T p492)

9. The Federal Highway Administration estimates that using computerized traffic control throughout the United States would save as many as 30 billion gallons of fuel per year. (F p492)

10. Some cities have reduced the cost of their automated traffic control systems by tying their traffic signals and computers into the cable TV networks already in place. (T p492)

11. One goal of future traffic control systems is to prevent gridlocks from developing. (T p493)

Emergency Service

12. Emergency road service dispatching is an excellent database application. (T p493)

13. Today's computerized emergency road service dispatching systems are excellent examples of decision support systems. (F p493)

Driving Simulators

14. Driving simulators are built into today's cars to for the convenience of the driver. (F p493)

Computers and Air Travel

Air Traffic Control

15. The Federal Aviation Administration (FAA) is responsible for monitoring and controlling the commercial, military, and private air traffic in the United States. (T p494)

16. There are approximately 20,000 aircraft in the United States. (F p494)

17. The Discrete Address Beacon System will use an air-based approach to aircraft collision avoidance. (F p494)

18. With the DABS system, every aircraft will carry a transformer--a device that sends a reply automatically when it receives a certain signal. (F p494)

19. The ground-based component of the BCAS system is called ATARS. (F p494)

20. The Beacon Collision Avoidance System is primarily an air-based system that can operate effectively even if ground-based systems fail. (T p494)

21. The Airline Pilots Association favors the BCAS approach over the DABS approach. (T p494)

The Holding Pattern

22. The most advanced computer systems available today are designed to completely eliminate the need for planes to wait in holding patterns. (T p495)

The Airborne Electronic Terrain Map System

23. The Airborne Electronic Terrain Map System is an advanced flight simulator used to train pilots. (F p495)

24. One way to accumulate electronic maps would be to use satellites; some of these can see objects the size of an orange on the ground. (T p496)

Flight Simulators

25. Because of their high cost, flight simulators are used to train pilots only for larger aircraft. (F p496)

Fuel Planning

26. Computerized fuel planning systems are emergency systems used only when there is an energy crisis. (F p496)

Computerized Reservations Systems

27. Computerized reservations and ticketing systems are generally considered to be DB/DC systems. (T p496)

Computers and Trains

Canada'a Intermediate Capacity Transit System

28. Conductors for ICTS trains are trained on state-of-the-art computerized simulators. (F p498)

29. ICTS trains use magnetic levitation. (F p498)

30. ICTS trains use LIMs. (T p498)

31. As ICTS trains brake, electrical energy is absorbed by the system rather than going to waste, as with conventional braking systems. (F p498)

32. With ICTS, all three computers must agree on the instruction to be performed before the train will obey; this is called a voting scheme. (T p498)

33. The ICTS computers are so precise that they can stop the train within inches of a designated spot. (T p499)

The Westinghouse People Mover System

34. The Westinghouse People Mover System at the Atlanta Airport safely transports approximately 1,000,000 passengers per day between the main terminal and the concourses. (F p499)

35. The computers that control the Westinghouse People Mover System at the Atlanta Airport can shut down the entire system in seconds. (T p499)

Magnetic Levitation Trains

36. The Planetran, the brainchild of Robert Salter at the Rand Corporation, may send trains through underground tubes at 6000 miles per hour. (T p499)

37. Planetran would operate via both LIMs and maglev. (T p499)

38. The Planetran would travel three times faster than the Concorde SST. (T p499)

39. The Planetran could travel around the world in less than two hours. (F p499)

40. In Japan, maglev trains are already transporting customers comfortably at more than 3000 mph. (F p499)

41. Planetran trains would float on air. (F p499)

42. Planetran would be so efficient that it would require only a few dollars worth of electricity to transport a passanger coast to coast. (T p499)

"Tracking" Railroad Cars

43. Computerized tracking systems are designed to determine where railroad cars are at any given time. (T p501)

Train Simulators

44. The Train Dynamics Analyzer is a computer system whose primary purpose is determining the stresses a train can handle as it climbs mountain ranges and goes around dangerous curves. (F p501)

Computers and Ships

45. The shipping industry has renewed its interest in sailing vessels. (T p501)

46. The St. Lawrence Seaway's Iroquois Lock handles more than 50,000 ship transits per year. (F p503)

Fill-in-the-Blanks

Computers and Automobiles

1. The system offered by General Motors that monitors and controls many engine functions is called the **Computer Command Control**.

2. A General Motors innovation is the microprocessor-controlled **modulated displacement** system, which automatically reduces the number of cylinders in operation as a car reaches cruising speed.

Car Pooling

3. Computer systems facilitate **car pooling** in which people share automobiles to reduce commuting costs and conserve fuel.

Traffic Control

4. Computers are used to synchronize traffic lights in computerized **traffic control** systems.

5. **Gridlock** occurs when traffic is so congested that intersections become blocked and impassable.

Driving Simulators

6. Computerized devices that are used in training courses for driving, flying, and sailing are called **simulators**.

Computers and Air Travel

Air Traffic Control

7. The **Federal Aviation Administration (FAA)** is the government agency responsible for monitoring and controlling the commercial, military, and private air traffic in American skies.

8. The FAA is working with private contractors to develop the next generation of computer-based air traffic control and **collision** avoidance systems.

9. The **Discrete Address Beacon System (DABS)** uses ground-based radar and computer systems to track each aircraft individually.

10. The device located on each aircraft that automatically sends a reply when it receives a

certain signal is called a **transponder** .

11. The **Automatic Traffic Advisory and Resolution Service** (ATARS) computers will work in conjunction with the DABS system.

12. The **Beacon Collision Avoidance System (BCAS)** , installed directly in each aircraft, interrogate the transponders in nearby planes to get their positions.

13. The collision avoidance system that is favored by the Airline Pilots Association is **BCAS**.

The Holding Pattern

14. Computers may eventually eliminate the need for **holding patterns** , so planes will not have to circle airports while waiting to land.

The Airborne Electronic Terrain Map System

15. The system currently under development by the Air Force, the **Airborne Electronic Terrain Map** System, will help a pilots fly safely at night, in poor weather conditions, or over unfamiliar territory.

16. Satellites can be used to gather information around the world to produce **electronic maps** for use by airline pilots with the Airborne Electronic Terrain Map System.

Flight simulators

17. Airline pilots use flight **simulators** to help learn how to fly new planes without leaving the ground.

Fuel Planning

18. During severe fuel shortages, **fuel** planning systems become particularly critical to the continuous operation of the airlines.

Computerized Reservations Systems

19. Computerized **reservations and ticketing** systems allow people to book airline flights, hotel rooms, and rental cars.

20. Reservations computer systems are of the type called **database/data communications (DB/DC)** systems, because they maintain massive databases of facts about upcoming

flights, and because they are normally tied into computer networks.

21. **Automatic ticketing** machines are microprocessor-controlled devices that are capable of printing airline tickets.

Computers and Trains

Canada'a Intermediate Capacity Transit System

22. The completely computer-controlled train system being developed in Canada is the **Intermediate Capacity Transit System (ICTS)**.

23. The trains of the ICTS system use **linear induction motors (LIMs)** that work on the familiar principle that opposite magnetic poles attract.

24. In areas where high reliability is needed, several computers can be used in a **voting scheme** where two or more of the computers must agree before an action takes place.

The Westinghouse People Mover System

25. The transit system used at the Atlanta Airport and built by Westinghouse is the **People Mover System**.

Magnetic Levitation Trains

26. The "train of the future" that is the brainchild of Robert Salter is called the **Planetran**.

27. The **Planetran** is a train of the future that will be able to travel from coast to coast faster than a jet plane.

28. The Planetran will use LIMs and magnetic **levitation** at speeds of 6000 miles per hour.

29. Magnetic levitation is also called **maglev**.

"Tracking" Railroad Cars

30. Railroad cars are monitored by computerized **tracking systems** that read the markings located on the sides of each railroad car.

Train Simulators

31. The train simulator used by Conrail to train engineers is called the **Train Dynamics Analyzer (TDA)**.

Computers and Ships

32. **Computerized sailing ships** take advantage of "free" energy and may well be the ships of the future.

A 5
B 6
C 7
D 8
E 9
F 0
G 1
H 2
I 3

Printed in Dunstable, United Kingdom

91342998R00255